EYE OF THE SHOAL

Also available in the Bloomsbury Sigma series:

Sex on Earth by Jules Howard
p53: The Gene that Cracked the Cancer Code by Sue Armstrong
Atoms Under the Floorboards by Chris Woodford
Spirals in Time by Helen Scales
Chilled by Tom Jackson
A is for Arsenic by Kathryn Harkup
Breaking the Chains of Gravity by Amy Shira Teitel
Suspicious Minds by Rob Brotherton
Herding Hemingway's Cats by Kat Arney
Electronic Dreams by Tom Lean
Sorting the Beef from the Bull by Richard Evershed and Nicola Temple
Death on Earth by Jules Howard
The Tyrannosaur Chronicles by David Hone
Soccermatics by David Sumpter
Big Data by Timandra Harkness
Goldilocks and the Water Bears by Louisa Preston
Science and the City by Laurie Winkless
Bring Back the King by Helen Pilcher
Furry Logic by Matin Durrani and Liz Kalaugher
Built on Bones by Brenna Hassett
My European Family by Karin Bojs
4th Rock from the Sun by Nicky Jenner
Patient H69 by Vanessa Potter
Catching Breath by Kathryn Lougheed
PIG/PORK by Pía Spry-Marqués
The Planet Factory by Elizabeth Tasker
Wonders Beyond Numbers by Johnny Ball
Immune by Catherine Carver
I, Mammal by Liam Drew
Reinventing the Wheel by Bronwen and Francis Percival
Making the Monster by Kathryn Harkup
Best Before by Nicola Temple
Catching Stardust by Natalie Starkey
Seeds of Science by Mark Lynas
Outnumbered by David Sumpter

EYE OF THE SHOAL

A FISHWATCHER'S GUIDE TO LIFE, THE OCEANS AND EVERYTHING

Helen Scales

B L O O M S B U R Y   S I G M A
LONDON • OXFORD • NEW YORK • NEW DELHI • SYDNEY

Bloomsbury Sigma
Bloomsbury Publishing Plc
50 Bedford Square, London, WC1B 3DP, UK

BLOOMSBURY, BLOOMSBURY SIGMA and the Bloomsbury Sigma logo
are trademarks of Bloomsbury Publishing Plc

First published in the UK in 2018

Photo credits:
P. 32: Image from the Biodiversity Heritage Library. Digitised by Harvard University, Museum of Comparative Zoology, Ernst Mayr Library, www.biodiversitylibrary.org. P. 33: Image from the Biodiversity Heritage Library. Digitised by the Smithsonian Libraries, www.biodiversitylibrary.org. P. 106: Image from the Biodiversity Heritage Library. Digitised by Harvard University, Museum of Comparative Zoology, Ernst Mayr Library, www.biodiversitylibrary.org.

A catalogue record for this book is available from the British Library

Library of Congress Cataloguing-in-Publication data has been applied for.

ISBN HB: 978-1-4729-3684-4
TPB: 978-1-4729-3681-3
eBook: 978-1-4729-3683-7

2 4 6 8 10 9 7 5 3 1

Illustrations on p. 51, p. 103 and p. 231 by Marc Dando

Bloomsbury Sigma, Book Thirty-five

Typeset by Deanta Global Publishing Services, Chennai, India
Printed and bound in Great Britain by CPI Group (UK) Ltd, Croydon CR0 4YY

MIX
Paper from
responsible sources
FSC
www.fsc.org
FSC® C020471

For Celia and Peter, Katie and Maddie,
with memories of Ningaloo.

Contents

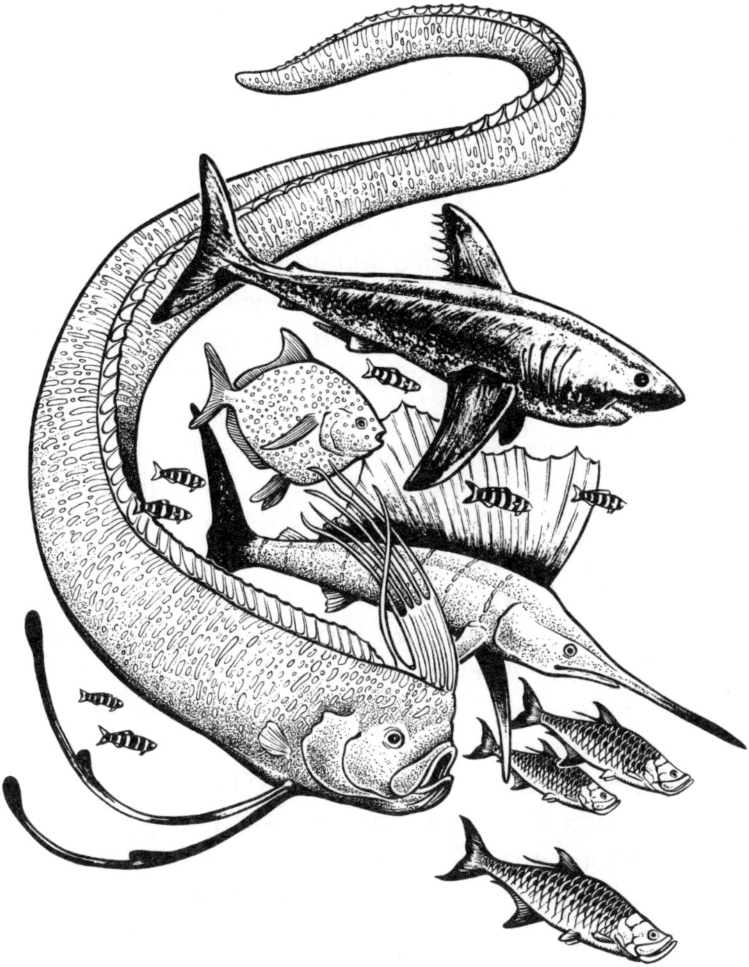

PROLOGUE

The wandering ichthyologist

Daylight fades, and a shoal of fish settles down for the night in a quiet pool in the Amazon rainforest. The fish are small, no bigger than your thumb to the knuckle, with forked tails, a golden body stripe and a red streak above each eye. They hide among the submerged, mouldering leaves that fell from the canopy high above, and they've all been tricked by a subtle message: *this is not a fish*. A leaf drifts towards the shoal. Like all the other leaves, it's brown and blotched; it even has a stalk at one end, where it was once apparently attached to a tree. But then, in an instant, jaws fling open and a huge mouth gulps and swallows a member of the oblivious shoal. Half a second later the Amazon Leaf-fish is once again a leaf.

Elsewhere in Amazonia, a male Splash Tetra with large pearly scales and red-tipped fins hangs patiently under a promising frond of vegetation, waiting for a female to show up and, hopefully, choose him as a mate. If she does, the pair will leap from the water and stick themselves to a leaf, clinging with suckers on their fins. The female will lay eggs, a dozen or so at a time, and the male will fertilise them with a squirt of sperm. Then they will both tumble back into the water. Splash Tetras keep doing this, jumping and falling again and again, until they've laid and fertilised at least 200 eggs in a dense clutch. Then the exhausted female will swim off, leaving her babies stuck to their leaf, out of reach of most aquatic egg-eaters and under the watchful eye of their father. It's up to him to make sure the growing babies don't dry out and once every minute he splashes them with a swish of his tail. Light bends as it crosses the boundary between air and water and this unwitting physicist adjusts his water jet accordingly, aiming just away from where he sees the eggs to make sure he hits his target. After two days the eggs will hatch, and the fry will drop into the water and swim away.

If they're unlucky, the new hatchlings might be spotted by a hungry Four-eyed Fish, *Anableps*. In reality, these hunters only have two eyes, perched high on their head like a frog's, but each is divided horizontally with two separate corneas and pupils. A single lens is flattened, like a human eye, on top and curved underneath, like in most other fish's eyes. With pale, elongated bodies *Anableps* spend their time lying just below the surface, gazing into two worlds. The lower half of each eye focuses down through the water to watch for other predators, while the upper halves stick out to scan the air and the water's edge for insects, or young tetras, that might fall in and become a meal for *Anableps*.

These Amazonian oddities are not the only unusual fish, picked from an otherwise monotonous crowd. The world's waters – fresh and salty, shallow and deep – are teeming with remarkable species.

In East Africa, in Lake Tanganyika, female cichlids* use their mouths as brood chambers. When a pair meets to spawn, the female lays eggs, the male adds his sperm, then the female sucks the whole lot into her mouth where the fertilised eggs will hatch and grow until they're ready for the outside world. Unless, that is, there's a Cuckoo Catfish nearby. These whiskery, white fish covered in black spots have behaviour just as duplicitous as their feathered namesakes. They swoop in while cichlids are laying eggs and add their own to the mix. Inside the duped female's mouth, the catfish hatch and make more space for themselves by eating all the young cichlids.

Further offshore from continental Africa, on the island of Madagascar, in deep, underground caves live gobies, less

---

* Pronounced 'sick-lid', in case, like me, you get ludicrously annoyed when you read a word and don't know what it sounds like.

than a centimetre (half an inch) long, pallid pink and with no eyes. Similar pale, eyeless gobies live nearly 7,000km (4,350 miles) away on the other side of the Indian Ocean, underground in a desert in Western Australia. Recent genetic studies revealed these two groups are closely related; they are evolutionary sisters. For these cave-dwelling fish, long-distance dispersal is really not an option. They're confined to their caves and can't risk emerging into daylight; they have no eyes to spot predators and no skin pigments to protect them from ultraviolet rays. The only likely explanation for their disconnected geographies is the movement of continents. A common ancestor to these two groups lived on an ancient, southern supercontinent. Then a split formed between Australia and Madagascar and for the last 100 million years or so the two cave systems, and their divided fish, have been slowly drifting apart.

And in the ocean's twilight zone 1,000m (3,300ft) down, where sunshine runs out, live some of the strangest, not to mention the most abundant fish of all. There swim small sharks with glowing spines on their back, probably to warn intruders not to take a bite; others have a pocket on each side of their head that holds glowing slime (and no one really knows why). This is also the territory of bristlemouth fish and lanternfish, sharp-toothed creatures that would fit in the palm of your hand. They illuminate their bellies with blue light and disguise their silhouettes from predators passing below, and some of them talk to each other with coded flashes of light, like fireflies. Deep-sea surveys show these two fish groups reign over the twilight zone. Together there are thought to be hundreds of trillions, maybe even thousands of trillions of bristlemouths and lanternfish alive today, far more than any other vertebrates (followed by 24 billion domestic chickens). Between dusk and dawn, great herds of lanternfish and bristlemouths leave the twilight zone and swim hundreds of metres towards the surface following their food, the wriggling masses of plankton that rise each night. It's the greatest animal migration on the planet and it takes place like clockwork on a daily basis, up and down again, all across the oceans.

Fish are one of the greatest success stories of life on Earth. They dominate the oceans and freshwaters that cover more than seven-tenths of the globe's surface. Consider the depth of the oceans, almost 4km (12,000ft) on average, and this amounts to somewhere between 90 and 99 per cent of all the available living space – the teeming, scrambling biosphere. Fish have ruled this colossal watery realm for hundreds of millions of years. Power has shifted with time between different groups, but the fish have always been there[*].

Precisely what defines fish and separates them from other animals is not, as we'll see in the first chapter, an entirely straightforward matter. Broadly speaking, fish are aquatic animals that breathe water through gills and have backbones, but with various notable exceptions. Setting that to one side for now, what is clear is that fish are by far the most abundant and also the most diverse of the vertebrates. Half of all the animal species with backbones are fish of some sort. There are roughly 30,000 fish species, and a similar total number of birds, amphibians, reptiles and mammals put together.

Fish range in size from 20m (65ft) Whale Sharks to 8mm (0.3in) tiddlers[†], and they come in a multitude of shapes;

---

[*] The normal plural of fish is fish. The older term, fishes, is the scientific norm referring to a group with multiple species, but I find that cumbersome. So, with apologies to the purists, I've chosen to use one term throughout this book for both one and many fish.

[†] *Paedocypris progenetica* from the peat swamps of Indonesia was thought to be the world's smallest vertebrate until 2012, when a newly discovered Papua New Guinean frog beat it to this diminutive record. 8mm is roughly the width of your little fingernail.

they can be serpentine ropes or round balloons, bullets or torpedoes, flat pancakes or cubes. Some fish are bright and kaleidoscopic, many are silver or sand-beige, others you can see right through; some are fast, some don't move at all; some live for weeks, others for centuries; some live in caves and no longer need their eyes and some drift around pretending they're leaves. Compared to other, more conservative animals, fish are supremely flexible and adaptable, and they've evolved unique adaptations to inhabit their liquid world. There's no single way to be a fish.

Yet so much of the brilliance of fish goes unseen and unknown. They live hidden beneath the waves, beyond the horizon. The shifting, tide-swept boundary on shores and riverbanks forms a dividing line between wet and dry, and between their world and ours. Since antiquity only the brave or the incurably curious have voluntarily crossed this line.

For thousands of years, people have hauled fish out of the water and brought them into the human world in two main ways. First and foremost, fish are food. Catching fish to eat is so deeply ingrained that we fish for fish, but we don't deer for deer or boar for wild boar (although some people do go rabbiting). Hunting for wild fish is an ancient practice. In a cave on the Japanese island of Okinawa, archaeologists unearthed fishhooks that were made from seashells at least 30,000 years ago. Chemical analysis of a 40,000-year-old human skeleton found near Beijing showed that this early human had eaten a lot of fish from rivers and lakes.

Today, global fisheries catch between approximately one and three trillion fish each year. This provides a primary source of protein for around a third of the global human population. For fishers, especially in small-scale fisheries, there's still a profound connection to the lives of fish. But for the majority of consumers, especially in high-income nations, there's a growing disconnect between the food we eat and where it comes from. Almost one in five young children in the UK think fish fingers are made of chicken. By the time most people come into contact with a fish it's already long dead, the head, fins, organs and bones are gone and the remains wrapped neatly in plastic or sealed in a tin. In the same way that a steak doesn't comfortably call to mind a mooing, cud-chewing cow, those flaky chunks of white and pink meat are almost impossible to relate to a wild, living animal. But the detachment is even more extreme for fish. We all know what cows look like, yet the appearance of many fish remains unfamiliar. In Britain people eat 70,000 tonnes of Atlantic Cod every year – around 1kg (2.2lb) per person – but only one in three seafood consumers can recognise these 2m-(6ft 6in) long fish, far longer than your outstretched arms, that are covered in gleaming bronze spots with a white goatee beard dangling from their chin. Fewer than one in five British consumers recognise a blotchy flattened fish, with two eyes looking upwards and a twisted mouth (a sole) or silver, bullet-shaped fish with big, wide mouths (anchovies). And those are among the most popular fish on people's plates. What hope of recognition is there for the lesser-known varieties that sometimes appear on the menu? There's the John Dory with its Mohican of spines, marbled copper skin and a pair of large, gold-rimmed spots, and the gurnard with its scarlet body and three 'fingers' on each side, which feel for food on the seabed.

Beyond the fish that we eat are the ones that swim into the human world through myths and folklore. Fish stories in cultures across the world tell of the deep-rooted and often conflicting feelings people have about these inhabitants of the depths. Mythical fish can bring their human companions good fortune, prosperity, renewal and knowledge. But they can also be fickle and dangerous, as shape-shifting demons that unleash floods, storms and earthquakes. Gods, goddesses and their entourage take the form of fish or swap their legs for a tail, sometimes willingly, sometimes as punishment. The original versions of mermaid stories in many countries are often uncomfortable and dark: outcast women escape underwater and transform into mermaids, then torment and curse the human world that banished them, luring people to their deaths. Hans Christian Andersen's *Little Mermaid* was so desperate not to be half-fish any more that she agreed to have her tongue cut out, and every step with her new feet felt as if she was walking on broken glass.

Many of these stories reflect the psychological barriers that make fish difficult to know or to like and, certainly, to empathise with. Fish seem to lack any emotions that we can interpret and understand, no smiles etched on their lips; just fixed, grumpy pouts. And place your hand on a living fish and it probably feels as cold as if it were lying dead on a supermarket counter; that just doesn't seem right for something that could still get up and rush away (although, as we'll see, not all fish are indeed fully cold-blooded, or ectothermic). I know several people who refuse to swim in the sea for fear of a cold, slimy fish brushing past. The best way to get over that fear is not to let those imagined fish swim by unseen, but to stick your head into the water and watch them instead.

This book is an underwater journey through the lives of fish. It's an exploration of what fish are and the things they do in their cryptic world. I'll unwrap fish from mysterious stories – and recount some of those tales – and I will unhitch them from their reputation as cold-hearted, unknowable beasts and present them as they truly are, the most captivating wildlife you can discover, get to know and admire wherever you are in the world.

Once we've settled the question of whether there are such things as fish (in chapter one) and taken a tour of their spectacular diversity (chapter two), each chapter will then explore a particular characteristic that helps fish to be so tremendously successful and abundant. We'll watch how fish move, how they gather food and how they avoid becoming someone else's lunch; we'll hear them sing and talk to each other, and see how they use light and colour to send out messages and to hide. Many of these attributes and behaviours are unique to fish and they combine to make them masters of the aquatic realm.

Now is the perfect time to rethink fish and get to know them better. For one thing, the sum of knowledge about this group of animals has never been greater. Armed with new tools and new ways of looking, researchers are making remarkable new insights; they're dispatching remote-controlled robots to spy on the deepest denizens, using molecular tools to decipher relationships and trace connections, and deploying miniaturised tracking devices to follow fish on journeys across entire oceans.

But fish are also collectively suffering from an onslaught of human impacts like never before. *The Sea Around Us* project at the University of British Columbia in Canada recently estimated that in 1996 the world went past the point of 'peak fish'. Until then, fisheries around the world were catching ever more wild fish year-on-year; more fishers with bigger boats, new fishing gears and technologies were venturing out and pulling a growing mass from oceans, lakes, seas and rivers. From 1997 onwards, however, the total catch began to steadily and significantly decline by around two per cent each year. And that's not because people are fishing less, but because the fish are running out. At a global scale, fisheries have taken too many fish. Wild populations are no longer resilient – they are no longer bouncing back like they used to.

The problems of climate change, as the seas become warmer and more acidic, and of chemical and plastic pollution add to the fish's worsening plight, making it all the more critical that we act now and don't let populations and species slip away unnoticed and unknown. While this book won't dig deep into the problems of overfishing or climate change or profess to offer detailed solutions, I hope by the end to convince you that fish matter, and that they're worthy of our attention and esteem. That strikes me as a good place to start.

On a brighter note, it could be that watching fish is good for you. In 2015, a research team tracked visitors at the National Marine Aquarium in Britain as they gazed through a huge acrylic window into a half million-litre (11,000-gallon) fish tank. The exhibit of a temperate reef, decorated with artificial kelp and sea fans, was in the process of being restocked. The researchers monitored visitors when the tank was empty, when it was partially stocked and then again when it was filled with fish. Their results showed that among a hundred visitors selected at random, the more fish they were looking at, the further their heart rate and blood pressure dropped. The study gently suggested that watching fish, even in artificial conditions, is relaxing and can soothe nerves.

The first time I watched fish in the wild I honestly didn't expect to be that interested. I was 15 years old and in southern California, a whole ocean and a continent away from home and the furthest I had ever travelled. It was our first family holiday outside Europe and a tremendous treat to be in this exotic country – the beds were so big I could share with my little sister and not get badly kicked, breakfast times involved choosing how tall I wanted my pile of pancakes, and we drove for hours in straight lines.

One thing I hoped desperately to see on that trip was a sea otter. I was a big fan. I'd watched the documentaries and got the calendars and T-shirts. Now it was time to see them for real. We drove up California's coastal highway, with the blue Pacific stretched out to the left and giant redwood forests towering on the right, and I kept pestering my parents to pull over so I could check if that dark dot in the water was a fuzzy marine mammal.

Dusk was approaching when we first spotted otters, not far offshore. They were floating in a raft*, busy wrapping themselves in kelp fronds so they could snooze in safety without drifting away on the tide. As they twirled round, contorting themselves to keep four paws dry, they twisted themselves deeper into my affections. I swear some of them fell asleep holding paws.

Perhaps because my ambition to see wild otters had been so easily and happily fulfilled, I was primed to contemplate something else in the ocean, something that wasn't so obviously adorable.

The following day, at a spot called China Cove a short way south of Monterey Bay, I stood on a high bluff and looked down into the clearest, most turquoise sea I had ever seen. It was a revelation to this Northeast Atlantic girl. Until then, the sea to me had meant losing my toes in foggy, green water as soon as I stepped in. And yet there I was, gazing through the water, watching a sea lion swimming tight circles. Stranger still was the fact that I could see what the sea lion was chasing.

A school of fish split neatly in two, then rejoined into a single, swirling constellation each time the sea lion rushed through. They might have been herring or sardines; it didn't occur to me to even wonder. As I watched, the sea lion occasionally cajoled a single fish away from the pack, then redoubled its efforts to catch up with it, flexing its body in nimble, rubber-skinned loops. And not once did I see the sea lion make a successful kill. Each time, the target fish found its way back to its companions and melted into the school.

This game of tag had me transfixed. There was something about those fish – being pursued but never quite caught – that got me hooked, and kept me watching.

* When they're in the water, a raft of otters is their collective noun. While scampering around on land they're known as a romp or a bevy.

There are lots of ways to watch fish. It's a pastime that doesn't have to involve getting wet. You can walk beside ponds, streams and rivers and peer at shadows and shapes, and lips rising to the surface to nibble food. In wellington boots, I've stepped out with the falling tide on the British coast and seen Tompot Blennies hiding under rocks with a red feathery tentacle above each eye, and found catsharks hiding among seaweed, newly hatched from their 'mermaid's purse' egg cases.*

A lot of people watch fish from the comfort of their own homes. The popularity of fish-keeping is at an all-time high. In Britain, one in ten households now has a fish tank. It's estimated that in the US more than a billion fish live in people's homes. Partly because they're generally kept in small shoals, rather than one or two at a time, there are more pet fish than there are cats, dogs, rabbits and hamsters put together. Maybe it's the growth in city living and busy lives – fish don't need to be taken for walks – but there's no doubt that fish are in fashion.

I've never kept fish. My main excuse is that I work from home and I know I'd never get anything done if there was a miniature, fish-filled ocean for me to gaze at. Then again, if you subscribe to the principles of *feng shui*, perhaps I could become more prosperous and productive if I carefully assembled an aquarium of nine goldfish, including a black one to absorb all the negative energy in the room.

My favourite way of watching fish is to get in with them. Scuba diving can be like going for a hike through a magnificent landscape, only with certain new rules. As a diver, you can't talk to anyone who's with you. All you can do is nod and gesticulate, point things out to each other and communicate in simple, pre-arranged hand signals – *Are you okay? Yeah, I'm okay.* This makes every dive a contemplative experience, a chance to get lost in your thoughts. Depending on where and when you dive, your

---

* Catsharks are also commonly known, confusingly, as dogfish.

view can be huge and expansive, or narrow and secluded. The water can be so clear it's almost not there at all, or it can be thick and soupy, shrinking your vision so the small, nearby things hold your attention.

Divers don't walk or run but drift and fly, often with very little effort required (but it's hard work if the aquatic breeze is blowing in the opposite direction to where you want to go). Imagine striding through a forest and being able to float up to explore the canopy high above, or that you could quite safely step off the edge of the Grand Canyon to see what the view is like from beyond the handrail, like a soaring hawk. As a diver you can effectively break free of gravity's pull. You can hover motionless in one spot and watch wildlife sweep past, or you can swim along with it.

Full immersion delivers a vivid sense of just how different life is in water, compared to our lives out on dry land, as hairy pedestrians inhabiting two dimensions. It's not only that fish can breathe water instead of air, but they skilfully contend with shifting currents, tides and waves. I will always be impressed, and not a little envious, of the fish's abilities to be utterly in command of their three dimensions. I've looked up at rolling waves and the silhouettes of churning shoals that are somehow confident they won't get pummelled on a reef's sharp crest (and they don't). I've watched schools of mackerel and herring race past in perfect formation. And I've watched in awe as a single fish holds itself, poised and almost motionless except for near-invisible tweaks and adjustments of its fins; a deft flick of its body and it darts up and backwards. I wish I could do that.

We can only ever be visitors to this other world, and usually only go there for an hour or so at a time. You have to keep checking your dive watch to make sure you don't stay down too long or go too deep. And as another reminder that you don't belong here, you can only breathe so much.

Sip through your tank of compressed air too quickly and your time below is curtailed.

But you don't have to scuba-dive to experience the fish's realm first hand. Often I leave my air tank behind and go down with just a lungful of air. Freediving, with no clanking dive-kit and noisy bubbles, lets me get closer to fish without scaring them away, but at most I can stay down for a minute or so. Simplest and easiest of all is to float on the surface, head down, with a snorkel tube to breathe through; then I can watch fish for as long as I want.

When I began writing this book I had the chance to go on a long journey. I went to places I've been before and others I've long dreamt of seeing, and I spent a lot of time watching and contemplating fish.

After 20 years away, I returned to Ningaloo, Australia's longest fringing coral reef that hugs 260km (160 miles) of the west coast. I was excited, but also worried to go back. This was 2016, a year filled with bad news for the environment, including coral reefs; a coral-bleaching epidemic was in full swing, triggered by warm water and killing large tracts of corals worldwide, including much of Australia's Great Barrier Reef.

Ningaloo had been one of the first reefs I'd got to know, and it impressed on me the raw beauty of wild places barely touched by human presence. This was where I first saw Humpback Whales leaping from the sea, and the shadowy humps of dugongs; I saw sea turtles almost every time I went in the water, and I swam with manta rays that flapped their black wings, far wider that I am tall. The reef itself was crowded with fish of every conceivable shape and colour, and across the lagoon were coral boulders, the bombies, like giant heads of cauliflower, that have been built over centuries by the coral colonies' thin skin of life.

I was doubly nervous, then: anxious that Ningaloo might not live up to my fond memories, and fearful that isolation hadn't been enough to protect it from the rigours of the modern world. But when I waded out and looked underwater I saw a scene still bursting with life, just as I remembered it – there in front of me were flickering clouds of metallic blue damselfish, parrotfish hurtling past, and yellow-striped snappers hunkered in a small cave; coral was growing in dense thickets, boulders and flat tables in browns, greens and pinks with no signs of bone-white bleaching; across the reef scarlet cuttlefish were fluttering their skirts and in the distance I glimpsed sleek reef sharks sliding past.

Whale Sharks had drawn me to Ningaloo all those years ago. I'd watched a documentary on TV about how they visit this region every year to gorge themselves on planktonic food stirred up when all the corals spawn on one dramatic night; coral colonies release eggs and sperm, which are eaten by little fish, which in turn are eaten by Whale Sharks. They're the biggest fish alive today, bigger than any other living animals besides the great whales. I decided I needed to see these giant, feasting fish for myself and wrote a letter to a local conservation group, asking if there was any possibility I could help them out – happily, there was. I was given a room in a shared house in return for my daily assistance on surveys of the Whale Sharks' movements and behaviour, as they cruised up and down the coast.

On my return to Ningaloo I leapt off a boat, the seabed 100m (300ft) below me, and squinted through the water as a gigantic shadow hove into view. There was the familiar blue-grey skin spattered with white spots; there was the entourage of hangers-on, the shoals of little fish riding the shark's bow wave and remoras, like fat snakes, clinging to its belly. The sharks I saw this time were mostly youngsters, a mere 5–6m (16–20ft) from head to tail. Perhaps they were the offspring of the ones I'd swum with on my previous visit to Ningaloo. No one is really sure how long Whale Sharks can live for, perhaps a century or more, but they certainly take their time to reach maturity – at least 20 years.

Elsewhere on my trip, I gathered fish encounters wherever and whenever I could. In Fiji, I stared at weedy rocks for long enough to spot the flicker of a stonefish's eye and see through its venomous camouflage; I had my fingernails picked clean by cleaner wrasse and I was chased by ferocious damselfish, a fraction of my size, defending their carefully tended seaweed gardens and trying to scare me off with a fusillade of warning drumbeats. In Aitutaki, in the South Pacific, when bad weather stirred up sediment and reduced visibility to a few centimetres, I clung to rocks and watched blennies smaller than a fingernail, poking their faces from small holes.

There were times I didn't expect to see fish, but there they were. Hundreds of miles from the coast, in Australia's arid interior, I climbed down into red gorges that cut through some of the oldest rocks on Earth. Streams spill through them, un-warmed by the sun, and collect in deep pools fringed by ferns and overlooked by trees festooned with squealing fruit bats. I dived into the green water to the bottom, 5m (16ft) down, and watched minnows darting through the weeds.

And many times I found myself inside a shoal, looking out with nothing but fish around me. There were shoals of many kinds. At dusk in Rarotonga, a mixed scrum of young fish stampeded past, feeding together across the reef flats – rabbitfish, parrotfish, surgeonfish – kicking up a plume of sand in their wake. Thousands of miles to the northeast, in Palau, I floated in open blue ocean surrounded by a shoal of hundreds of Red-toothed Triggerfish, flapping two fins like so many butterflies flying on their sides. When something spooked them, they all swooped to hide in holes in a vertical coral cliff. And off the Tutukaka coast of New Zealand, in the archway of a submerged cavern the size of a cathedral, I was encircled by a shoal of Blue Mao Mao. In outline these are archetypal fish, oval bodies with forked tails, and they are the colour of a clear summer sky. A congregation of identical faces turned to look at me before gently flowing around me as if I were a stone dropped into a living, shimmering river.

You may be wondering if I'm about to embark on a travelogue, documenting all the places I went and all the fish I saw on that journey. I did see a lot of fish and some will appear in the coming pages, joined by others from previous encounters, but a detailed fish log is not my plan. While it's always exciting to see animals I've not seen before, I'm not driven by a single-minded compulsion to find ever more species and check them off a list. That's not the point of watching fish.

The point is to explore and notice new things, to contemplate fish and the rest of the world in new ways. The lives of fish tell us not only about their mastery of oceans and freshwaters, they also uncover greater truths about life on Earth. They tell us about the possibilities of evolution. They reveal the inner workings of ecosystems, and they tell us about the adaptability and resilience of living things.

Spend time watching fish and a flood of questions follows: How do shoaling fish avoid bumping into each other? How do they avoid the jaws of fast-thinking predators? Why do fish scrawl secret graffiti across their bodies? (And who's reading it?) How do thousands of fish species get along when they live in crowded places, like the Great Lakes of Africa and the Amazon basin?

Why does a manta ray keep its brain warm? How do deadly pufferfish manage to not poison themselves? What do fish think about all day?

Why do little fish willingly climb inside the mouths of predators with only the power of dance to protect themselves? What do fish do when their water dries up? How did catfish learn to catch pigeons? How do fish cross entire oceans, then find the same stream where they were born? How did they strike a deal with bacteria so they can see in the dark?

This book will answer these questions and many more that arise as we submerge ourselves in the fish's water world. We'll meet the scientists who set out on great adventures to watch fish, to search for answers, to eavesdrop and gather more details of their lives. This book will tell the stories of those dedicated fishwatchers who make careful plans and often come back with unexpected discoveries. And we'll see that science isn't simply about listing answers to questions, but it's the whole exciting, arduous, creative and sometimes messy process of reaching for those answers, and understanding a little more about how the world works.

Along the way I hope to convince you that fish are worth finding and watching and paying more attention to. Unlike other sea creatures, like molluscs with their shells, fish don't leave elegant parts of themselves behind on beaches for us to pick up and admire. When fish die, their gemlike colours fade and their bodies soon decompose and decay; all that most of them leave behind are rotting carcasses and godawful smells. The best way to appreciate fish is to see them while they're alive.

Maybe you'll begin to see your pet goldfish in a new light. Maybe you'll go and stand by the tanks in your local aquarium and notice fish doing things you didn't see before. And maybe you'll take a chance to go and find some water and peer in, or better still to jump in and take a look, to see if there are fish, and to find out what they're up to. With 30,000 species and trillions of individuals, you aren't going to run out of fish to find and ponder with any time soon.

CHAPTER ONE:
ICHTHYO-CURIOSITIES

CHAPTER ONE

Ichthyo-curiosities

Until a couple of years ago, I hadn't given much thought to the matter of whether or not fish actually exist. I had just rather assumed that they do, having spent a good deal of time watching, following and studying them. Then someone asked a question that caught me off guard.

'Is it true that there's no such a thing as a fish?'

This was followed by uncomfortable, silent seconds that ticked by as I tried to think of something to say. It didn't help that there was a live microphone in front of me and 300 people watching and waiting for a knowledgeable and preferably witty answer.

I'd been invited to appear on the BBC panel show, *The Museum of Curiosity*, the radio-sibling of the television series *QI*. In each episode the museum's curator, John Lloyd, introduces a panel of three guests including someone who's done something genuinely noteworthy (astronaut Buzz Aldrin appeared on one show), plus a comedian and an academic or someone else suitably nerdy, which is where I came in.

Guests are asked to donate something to a vast, imaginary museum and then explain to the audience why they picked it. The donations are often quite silly, but also thought-provoking. By the time I joined in, the museum already had on display a yeti (donated by actor Brian Blessed), the Big Bang (from cosmologist Marcus Chown), the Icelandic Grímsvötn volcano (from music producer Brian Eno) and 'tempting fate' (from comedian Tim Minchin). For my donation, I requested that a large aquarium tank be installed in the museum and filled with seahorses, including some pregnant males. The fact that these seahorses were quite definitely a type of fish, in my mind at least, didn't help me compose an intelligent answer to John's question.

As I sat facing the studio audience, all looking at me like an attentive school of fish, my brain froze and I couldn't recall anything useful or especially interesting to say about how valid the term 'fish' really is.

I think that's what John was getting at: is there a biologically rigorous definition of a fish? Do such mismatched animals as goldfish and pufferfish, flounders and minnows, anglerfish and sunfish, really all belong together in the same group? Or are they just a random assortment of creatures that all happen to be good at swimming? Is there any more reason to group fish together than, say, spiders and octopuses because they all have eight legs?

Whatever I said it wasn't worth repeating, and the programme's editors thankfully chose to leave that bit out. The fish question, along with my dithering answer to it, didn't make the final cut.

Historically, the idea of what is and what is not a fish has been considerably shaky. Things did at least get off to a good start with Greek philosopher Aristotle's study *Historia Animalium*. His nine-volume work, composed back in the fourth century BC, was one of the earliest zoological texts. Among his observations were detailed insights into the lives of water-dwelling creatures, and they show how Aristotle was clearly convinced of the existence of fish.

He sorted all known animals into various groupings, firstly into blooded versus bloodless animals, the vertebrates and invertebrates respectively.* Within that dichotomy, he placed each animal in a separate group that he referred to as a genus. For the aquatic animals this included Malacostraca

*The majority of invertebrates do in fact have blood, but most lack a closed circulatory system, or vessels through which blood flows.

(crustaceans), Zoophyta (including jellyfish and sponges), Ostracoderma (bivalve molluscs) and fish, which he described as living in water, being 'destitute of feet' and instead having winglets or fins, usually four of them, but only two in the case of the 'lanky fish' like eels. The fish don't have hair, or feathers and they're usually, but not always, covered in scales; they have blood, like the four-footed land-dwellers; some give birth to live young and some lay eggs. Aristotle offers thorough accounts of fish anatomy, how they breed and how fishermen go about catching them. In all, he name-checked around 120 fish species, many of which he would have encountered in the Mediterranean and which still live there today, including mullet, moray eels, tuna and parrotfish.

Crucially, Aristotle separated fish from whales, porpoises and dolphins, which he placed in their own group – the cetaceans. Even though they are legless swimmers that occupy the aquatic realm, Aristotle saw that cetaceans share many similarities with land mammals, including breathing air through lungs and suckling young with milk. In his view, they were undoubtedly different from fish.

However, Aristotle's was the one of the last sensible accounts of aquatic creatures for a fair while, as the waters grew far murkier. For almost the next two millennia, accounts of aquatic creatures became mingled with tales of strange and imaginary beasts. In medieval times, books of animals were known as 'bestiaries'. These lavishly illustrated compendiums mainly included the retellings of ancient fables that instructed readers on how to live good lives. There were sirens with the upper body of a human, and fish from the waist down, who lured sailors to sleep with their songs, then attacked them; the moral of the story was to avoid the temptation of earthly pleasures. Some sea creatures bore a passing resemblance to real species such as the *serra*, depicted with enormous wings and leaping into the air, like a flying fish. *Serra* raced against ships, envious of their speed, but before long flopped beneath the waves once more. Such stories warned of people distracted by envy who, like the fish, will ultimately plunge into dark, hellish waters.

It wasn't until the Renaissance that lifelike sea creatures began to rear up from the monster-infested depths. In the middle of the 16th century, European scholars picked up Aristotle's ideas and once again started asking serious questions about aquatic animals. The notion of fish started to edge closer to reality.

In the Rare Books Room at the University Library in Cambridge I hand over a pile of flimsy pink paper slips, filled out in pencil, to the librarian. He checks my library card one more time, then wheels out a trolley laden with books. I carefully arrange them on a table, propping them up on grey cotton cushions. Together these tomes are worth tens of thousands of pounds, including original copies of the world's oldest books about fish, close to 500 years old. Throughout them all threads a trail of ideas about animals that live underwater.

The first three books I look at were published over the course of three years, in the 1550s, by a trio of European writers. These men led similar lives, training in medicine and travelling widely in Europe; they all knew each other and probably met at some point in Rome, and they all shared a great passion for aquatic life.

In 1553, Frenchman Pierre Belon published *De Aquatilibus* (Latin for 'aquatic animals'). It's a small, leather-bound volume, wider than it is tall, that sits in my hand like a flick-book, but I won't be doing any flicking. The narrow pages are filled with Latin text and illustrations of 110 aquatic animals. Some are realistic depictions – there are flying fish, tuna, eels and lampreys – while others are cartoonish versions of real fish. The pufferfish looks like a drawing made by someone who's never actually seen one

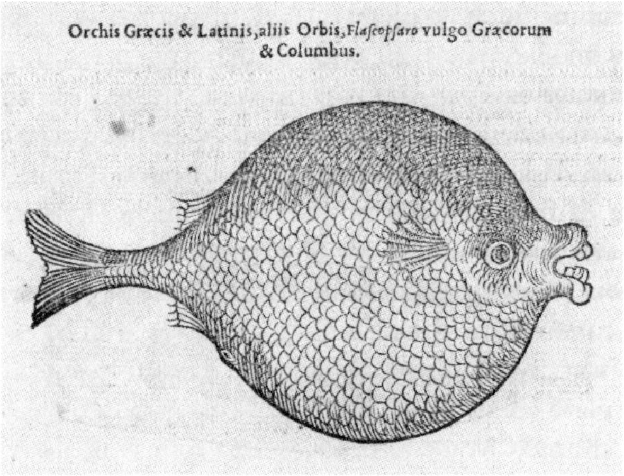
Orchis Græcis & Latinis,aliis Orbis,Flascopsaro vulgo Græcorum
& Columbus.

Pufferfish from Pierre Belon's De Aquatilibus, *1553.*

but has heard stories about fish that inflate like a balloon until they're as round as a full moon.

As I turn the pages it quickly becomes obvious that Belon subscribed to the theory that anything living in water must be a fish. He did, however, divide them into two groups. First were the bloodless fish, following Aristotle's idea, including octopus, seashells, sea urchins and crabs. Then there were the fish with blood, the tuna, sharks and so on, plus otters, dolphins, whales, water rats and beavers; there's even a hippopotamus trying its best to eat a poker-straight crocodile (presumably Belon saw these two species, or at least heard about them during his travels in Egypt, although who knows why his crocodile is so stiff). He added to this category various sea monsters that he probably picked up from medieval bestiaries, such as the fish bishop, an ecclesiastical gentleman walking around with a giant fish over his head.

Belon is often remembered as the founder of ichthyology, even though it was only a year later that two other fish books were published. In 1554, fellow Frenchman Guillaume Rondelet published *De Piscibus Marini* (Latin for 'The Fish of the Sea'). It describes more species than Belon's books, 244 in total, and includes a similar mix of Aristotelean and medieval ideas. Rondelet wrote about dolphins, crocodiles, all sorts of invertebrates and a handful of mythical beasts, like the lion covered in scales with a human face, although he remained suspicious, having never actually seen one for himself.

That same year saw the publication of another fish book, and the grandest so far. I untie the strings holding together the worn copy of *Aquatilium animalium historiae* ('The history of Aquatic Animals') by Italian scholar Hippolytus Salviani and I peer inside at the intricate copperplate engravings of some 96 varieties. Compared to the chunky woodcuts dispersed throughout the pages of Belon and Rondelet's works, the engravings simmer with life. An inflated pufferfish looks up at me and I can just imagine what it would feel like to pick it up in two hands; a moray eel slithers life-like across the page.

Together, Belon, Rondelet and Salviani breathed new life into the scientific study and appreciation of fish that had begun with Aristotle some 1,800 years previously. The issue still remained over exactly what fish are, and what sets them apart from other animals. Nevertheless, these books firmly established ichthyology as one of the oldest recognised branches of zoological study. And they remained widely influential until another work was published, more

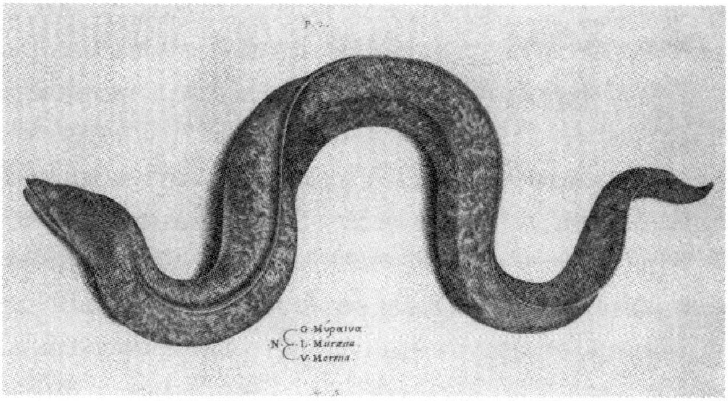

Moray eel from Hippolytus Salviani's Aquatilium animalium historiae*, 1554.*

than a century later, which became perhaps the most notorious of all fish books.

De Historia Piscium ('The History of Fishes') was published in 1686, some 14 years after the death of Francis Willughby, the British man whose name appears on the front cover. It was his friend and former tutor, John Ray, who completed much of the work. The two had met at Cambridge University, and several years later set off on a journey around Europe with a grand ambition of collecting material for a series of books that would reform natural history. Their work would be grounded in careful observations of real, living things. Ray devoted himself to the plants, Willughby to birds, insects and fish. Between 1663 and 1666, they travelled around England, France, Germany, the Netherlands and Italy, by boat, horseback and carriage. Both men gathered specimens as they went, buying books and paintings, watching over the shoulders of other scholars as they dissected animals and examined their insides.

Shortly after the pair returned to England, as they were preparing their collections, Willughby died of pleurisy (inflammation of the lungs), aged just 36. Ray stepped in and committed himself to finishing his friend's works. First was the bird book, *Ornithologiae Libri Tres* (Ornithology in three books), published in 1676. Ray then turned his attention to fish.

What began as Willughby's project came in time to be strongly shaped by Ray's ideas about the natural world. From the outset, Ray declared this wouldn't be an encyclopedia, gathering together every known fact (and fiction) about aquatic animals. That meant no more sirens or fish bishops. He adopted a fish definition in line with Aristotle's: animals that only survive in water that have no legs and no hair. This kept in cetaceans, because they don't do well on dry land, but left out hippos and crocodiles. Ray also made a significant step forward by excluding crabs, molluscs, jellyfish and all the other invertebrates. Possessing a spine was now an obligatory attribute for any animal to be considered a true fish.

Ray also worked hard to overcome what he called the 'multiplication of species'. Vague descriptions often led to a single animal being named several times in separate texts, artificially boosting the total number of described species and generally confusing things. This problem, known as synonymy, still persists (the most over-named species in the world today is *Littorina saxatilis*, a type of North Atlantic sea snail, which has been named at least 128 times). Ray planned to clamp down on such sloppiness by focusing on external attributes and markings, instructing readers so they could look at a fish and correctly identify it. This was the world's first definitive fish-spotting guide, listing 420 carefully selected species.

When it came to publishing *De Historia Piscium*, it was originally the Bishop of Oxford who pledged to privately finance the printing costs, on condition that the Royal Society in London bought a hundred copies. Eventually, though, the Royal Society opted to print the book at their own expense, a decision that almost saw the end of the scientific learned society, which at that time was less than 30 years old.*

Even though Ray considered the book ready for publication, some of the Royal Society's fellows spent an additional 10 months helping to revise the text, check fish names, make corrections and expand the work – in particular the illustrations. Ray had been undecided about whether or not to include drawings in the book at all, but he was eventually convinced that the best way for readers to identify their own fish was with the help of accurate

* The society's full title was originally 'The Royal Society of London for improving Natural Knowledge'.

pictures. It was too expensive to commission brand new drawings, so Ray and the other society fellows instead hunted through existing books to find the best images to copy. These were then cut out and given to a team of engravers, who produced 187 full-page plates.

Nevertheless these copied images were still costly to produce, and in order to finance the whole operation the Royal Society fellows and other leading academics were asked to subscribe to individual plates, at a cost of one pound per fish. In return, subscribers would get a discount on the finished book, and also have their name printed next to their species.

I open up the University Library's copy of *De Historia Piscium* and see lots of intricate fish pictures, and lots of famous names. There's a flying fish, clearly copied from Salviani's book, inscribed with the name S. Pepys. By the time this fish book was published in 1686, Samuel Pepys had finished writing the decade-long diary that made him famous. He was then president of the Royal Society and a generous contributor to the book; his name appears on 80 plates, including the hammerhead shark and 'blewe shark' (most probably a basking shark). Other fish-sponsors included the chemist Robert Boyle, the collector Hans Sloane and the architect Christopher Wren.

In total £163 of donations were raised, but this fell a long way short of the final cost. The Royal Society had to foot the remainder of the £360 bill, most of it to pay for the engravings. Five hundred copies were printed and put up for sale for roughly a pound each, more for those printed on better quality paper. The books should have easily made back the money, but the society had drastically overestimated the demand for Ray's lavish fish book and it was left with a lot of unsold copies. This spelt bad news for another scientific author hoping to publish that same year.

Laden with debt, the Royal Society was forced to withdraw their support for Isaac Newton's *Principia,* the monumental study laying down his universal theory of gravitation and how it could explain the motions of planets, comets and other heavenly bodies. Instead Newton's friend and Royal Society fellow, Edmund Halley, stepped in and stumped up the cash to publish the book, which went on to become a scientific milestone.

Edmund Halley is best remembered for correctly predicting the periodic return of the comet that now bears his name, but he also had plenty to do with fish. He spent several years captaining scientific expeditions around the world, sending illustrations and descriptions of exotic fish back to the Royal Society. He also invented a diving bell, which he wrote about in a journal article in 1714 entitled 'The art of living under water'. He tested it out for an hour and a half watching fish from the bottom of the River Thames. And in 1687, the year Newton's book was published, Halley was given 50 copies of *De Historia Piscium* by the Royal Society, instead of his £50 salary.

Even though it was a commercial flop, Ray and Willughby's fish book wasn't a total disaster. It showed that the Royal Society took the science of ichthyology very seriously. Not only did these eminent scientists invest a lot of money in it, but many fellows devoted considerable time to correcting and augmenting the book. The end product was a tome that made important advances in the careful naming of species, and took a major step towards a decisive picture of what fish are and where they fit on the tree of life.

Fifty years later another book of fish was published, with another dead man's name on the cover. Peter Artedi had grown up on the shores of the Gulf of Bothnia, at the northern tip of the Baltic Sea in Sweden. There he became fascinated by fish, and by the age of 11 was already studying and dissecting them, something he would continue doing for the rest of his regrettably short life.

At Sweden's University of Uppsala, Artedi studied medicine, still then the closest occupation to a scientific career. While he was there he became good friends with Carl Linnaeus, the man renowned today for revolutionising the study and naming of species. That was still to come when the two students spent time together in the late 1720s, sharing and developing ideas about the natural world.

In 1734, Artedi left Sweden and set off to further his medical education overseas, while at the same time expanding his studies of fish. He went to London to meet Hans Sloane and examine the fish specimens that formed part of the founding collections at the British Museum. Next Artedi went to the Netherlands, with plans to study for a doctorate in medicine, and it was there he caught up again with Linnaeus.

By now Artedi was penniless, and Linnaeus suggested he could earn some money working for Albertus Seba, a great collector who was compiling a catalogue of animal specimens brought back for him by sailors and merchants from the East and West Indies. These were to be the last fish Artedi saw and studied. In September 1735, after staying late with friends at Seba's house in Amsterdam, Artedi was on his way home in the early hours of the morning when he fell into a canal and drowned.

Linnaeus heard the news two days later and hurried to Amsterdam. He paid off the debts Artedi owed his landlord in exchange for five ichthyological manuscripts his late friend had been working on. Three years later, Linnaeus published the edited works in a single volume, *Ichthyologia*.*

---

* In his historical novel *The Curious Death of Peter Artedi*, marine scientist Theodore W. Pietsch reimagines the relationship between Artedi and Linnaeus as close friends, perhaps lovers. Linnaeus is depicted as a towering and merciless genius who plotted Artedi's

This is the next book I reach for in my pile at the University Library, a modest book, the size of a paperback novel with delicately marbled endpapers. It has no illustrations, just lists of names and descriptions. At the time, this was a fish book like no other.

Compared to his predecessors, Artedi did several novel things with his fish studies. He adopted John Ray's ambition of tidying up the profusion of fish names, but took the idea further. For years Artedi had pored over ichthyological texts, including all the books I've seen so far at the library. He teased out and matched common names in different languages, cross-referencing them to show how different writers in different countries were often talking about the same species.

Then he took those species and went to great lengths to sort them carefully into groups. First he divided the animal kingdom into classes, including one each for mammals, birds and fish. The fish class was then divided into five orders depending on whether they have bony or cartilaginous skeletons (the latter are the Chondropterygii), whether their fins are spiny (Acanthopterygii) or soft (Malacopterygii), and their gills exposed (Branchiostegi); cetaceans still feature as fish, in the order Plagiuri. These orders were in turn split into maniples, then genera, then again into species.*

In total, Artedi's *Ichthyologia* contains 52 genera and 242 species. This was as close as anyone had yet come to a definitive list of the world's fish that were known to western science. In particular, Artedi expanded the horizons of fish

---

murder in order to get his hands on the fish manuscripts and appropriate his rival's ideas for his own work. The accuracy of this version of events remains unclear.

* Biologists still today divide the living world along similar lines but with a few different names thrown in. The main hierarchical groups, nested within each other are: kingdom, phylum, class, order, family, genus (plural genera), species.

studies by including several species from beyond European shores. Exotic specimens in Sloane and Seba's collections that Artedi named included seven species of butterflyfish from tropical coral reefs, along with *Anableps*, the Four-Eyed fish from South America.

Artedi based his fish categories in part on the meticulous observation of real specimens; simply re-writing other people's work would not do, he had to see the fish for himself. He paid closer attention to detail than anyone had before in any branch of natural history, not just in ichthyology. He spent days working on a single specimen, counting fin spines, scales, teeth and vertebrae, and drawing the internal organs. His work was a glimpse of the painstaking science of taxonomy, the classification of species. While Artedi has been largely forgotten, Linnaeus is celebrated as the founder of modern taxonomy, even though his late friend had helped him develop many of the key ideas behind his new, orderly view of the living world. In his famous book, *Systema Naturae*, Linnaeus followed Artedi's idea of nested animal groupings, species within genera within families and so on. He also adopted all of the fish species described by Artedi, with a few notable exceptions.

Linnaeus was the first person to stop referring to cetaceans as a type of fish; he confidently placed them among the mammals. Artedi had obviously recognised that the whales, dolphins, manatees and narwhals were radically different from other aquatic vertebrates, but he couldn't quite shake off the historic habit of grouping together all the backboned swimmers. It was Linnaeus who finally did the decent thing and introduced the notion of marine mammals. This left him with almost 230 true fish, all of which are still recognised today. There are, though, far more living species of fish in the world than that.

Over the next century, scores of explorers and naturalists ventured across the globe with various ambitions in mind, including the finding and naming of species – many of them fish. From the rivers and lakes of North America and the frozen seas of Iceland and Siberia to the warm, clear waters of the Red Sea and Indian Ocean islands, thousands of new fish species emerged into the world of science.

As well as the adventurers who caught and preserved animals there were regiments of desk-bound compilers, most of them back in Europe, who eagerly wrote and re-wrote treatises describing the new specimens that poured in from distant shores. The most prolific of these describers was a man who isn't chiefly remembered today for his aquatic works, yet he wrote a series of books that ushered in a new era of ichthyology.

Back in the Rare Books Room, I've almost reached the bottom of my pile of fish books. I'm down to just three, picked from a monumental 24-volume series that was the biggest, most ambitious ichthyological work for a long time, perhaps ever. All in all, it took decades to compile, from 1828 to 1870, and still it outlived its authors, who died before completing this version of *Histoires Naturelles des Poissons* ('Natural Histories of Fish').

The first author was French zoologist Georges Cuvier, from the Muséum national d'Histoire naturelle in Paris. He was a pioneer of comparative anatomy, the study of physical similarities between different species; he was the first to suggest the earth was once ruled by giant reptiles, and persuaded a sceptical scientific establishment that species can and do go extinct. Cuvier also had the grand ambition of cataloguing all the known fish on the planet. To do this he amassed a fish collection like no other. He built up a network of travelling collaborators who collected fish all over the world and sent their pickled bodies back to Cuvier in Paris. There he set about the task of identifying, classifying and naming them all, and began writing his volumes of *Histoires*.

I open a volume filled with fish illustrations that look strikingly real. The etchings were hand-painted in full colour, and none would look out of place in a modern fish identification guide. Fish after fish I recognise; there are red squirrelfish with big round eyes, anemone fish with white bands across their orange bodies and magnificent swordfish and sailfish, spread out over two pages that fold out.

Cuvier wrote detailed descriptions of hundreds of fish in numerous volumes published up until his death in 1832, after which his student, Achille Valenciennes, continued alone. In all they jointly published 22 volumes, and in a leap forwards from Artedi's time less than a century earlier, they described 4,512 fish species. Yet this immense work still remained unfinished. In 1865 another French zoologist, Auguste Duméril, picked up where Cuvier and Valenciennes left off and released a further two volumes, devoted to sturgeons, garfish and sharks. These were the final entries in the most complete set of fish catalogues ever made, and they gave an inkling that the search for fish species had really only just begun.

I leave the library's Rare Books Room and climb the narrow metal stairs, passing through 150 years of history back to the present day, to the open book shelves where the less precious volumes are kept. Here I search for the newest books about fish.

The 600-page *Fishes of the World* by American ichthyologist Joseph S. Nelson is one of the latest. The book doesn't contain a list of individual species – there's far too many for that. It gives a total figure of 27,977 fish that are known and named, along with a rough estimate that there are at least 32,500 when including all the ones that scientists are still to find.* I flip through the pages, mainly looking at the pictures. Simple black and white line drawings depict the major fish groups (chiefly orders and families) and it's plain to see they all have an essential fishiness about them. Admittedly they're a eclectic bunch, but most share key characteristics – a pair of eyes, a mouth and a set of fins, which shift position and shape from group to group. The eyes can be tiny dots or great round orbs, and are occasionally gone altogether. The mouth can be small and puckered, huge and gaping, or underslung and frowning. And the fins generally include a tail at the back, a pair of pectoral fins, one on each side, a dorsal fin on top, a pair of pelvic fins underneath and, further back, an anal fin. The fins can be neat triangles or rounded fans or long, trailing ribbons; tails can be sharp forks or smooth crescents.

There's also a key addition to this book of fish that I didn't see in any of the older tomes. *Fishes of the World* contains a large diagram of interconnecting lines and names. A fish evolutionary tree.

Similar to family trees that map out human ancestry, evolutionary trees show how life forms are related to each other. They tell a similar story of descent and present an idea of what happened over millions of years as species evolved, one from another. This fish evolutionary tree is what I wish I'd thought of back in the BBC Radio Theatre. Is there such a thing as a fish? Yes. Absolutely there is. And an evolutionary tree helps tell us why.

Put simply, all the animals we know of as fish are related to each other, and they all appear on the same part of the tree of life. The fish aren't scattered through the tree's

* Other publications quote various numbers for the known fish species, but all go for about 30,000. Also, you might be wondering how it's possible to estimate the number of unknown species, but there are techniques for making such predictions based on the patterns of what's already known.

crown. We could saw off just one branch and that would contain them all.

Originally, diagrams like this were based on the appearance of living things, by comparing the similarities and differences between them. Ones that look most alike, both inside and out, are more closely related to those that look very different. This returns us to Aristotle's idea that fish are all the animals that live in water and have a backbone, gills and limbs in the form of fins. To these we can add other fish features (many of which we'll see through the rest of this book) that aren't found in any other animals.

To clinch this notion that all fish are related we now have genetic studies. Take a piece of any fish, run it through a DNA sequencing machine, and the results will come back showing that it shares much of its genetic code with other fish, because they inherited that code from a common ancestor. Fish are more similar to each other, genetically speaking, than to other living creatures, including jellyfish and starfish (the 'fish' in these invertebrates' names really is misleading; they have no backbone, and are definitely not fish).

There does, though, remain one niggling complication. It's a snag that leads us right back to where we started, and the issue of whether fish are simply all those creatures that swim. Because part-way down the fish evolutionary tree is a branch labelled Tetrapoda – the backboned, land-dwelling animals with four limbs. This includes crocodiles and hippos and various other creatures that were thrown out of early fish books. And yet here they are still hanging around.

If we grab that fish-bearing branch of the tree of life that we just sawed off and shake it, a lot of other animals will tumble out: not just fish, but toads, bats, rattlesnakes, pelicans, giraffes, polar bears and people. We're all interlopers among the fish, along with the rest of our mammalian, amphibian, avian and reptilian relatives – all the tetrapods.

This, ultimately, is why discerning biologists grumble about the term 'fish'. Fish are not what's known as a monophyletic group, an arrangement of species that we see, for example, in molluscs; you can make a single cut on the tree of life and have only snails, squid, octopus and other molluscs. The same goes for insects and flowering plants, but not for fish. The problem is that to have only fish on your branches, you need to make a second cut and chop out the tetrapods. This makes fish paraphyletic. And as far as those grumbling biologists are concerned, you can't go around pruning off branches and hope no one will notice. Such a tactic, they say, is artificial and arbitrary.

The paraphyletic fish aren't as artificial or arbitrary as *poly*phyletic groups. These are dotted here and there across the tree of life. To gather them together takes multiple snips and cuts all over the place. Like the pachyderms: elephants, rhinos and hippos are not at all closely related within the mammals, they just all happen to have thick skin. Admittedly though, most polyphyletic groups are accidental. As soon as someone notices the mistake, often via genetic studies, the offending species are quickly disbanded and packed off to their correct groups.

Besides fish, a few other paraphyletic groupings also remain in common use. Dinosaurs suffer under a similar misnomer because within their midsts the birds evolved. For the sake of biological accuracy, we should refer to birds as dinosaurs or, if you prefer, you can call the likes of *T. rex* and *Stegosaurus* 'non-avian dinosaurs'. By the same token, the fish could be 'non-tetrapod vertebrates'. But generally we seem to be getting along just fine with birds, dinosaurs and fish.

Strictly speaking, if we insist on sticking to the rules of taxonomy, humans are fish together with all the other land-dwelling tetrapods. But that's not a very helpful view. In *Fishes of the World*, Nelson was content to point out this polyphyletic anomaly and stick with a simple definition of fish as being all those aquatic vertebrates that have gills throughout their lives, and limbs in the shape of fins. As he pointed out, the term fish is, quite frankly, convenient. All we need do is come to terms with the fact that fish gave rise to tetrapods; some fish evolved to live permanently out of water and ultimately evolved into amphibians, then reptiles, birds and mammals. Other fish stayed behind and carried on evolving and doing their own things underwater. To see how and when that happened we need to take a closer look at the fish's branch of the tree of life, and at the same time we'll get a good idea of what exactly it means to be a fish.

Sedna the sea goddess

Inuit, traditional

In a village in the far north there was a woman called Sedna whom many men wanted to marry. She rejected them all until a hunter came from a faraway island. He promised her the finest food and furs, so she agreed to marry him. When they were alone on his island, Sedna's husband revealed he was not in fact a man, but a bird. Sedna was outraged, but she was trapped. When her father heard what happened he came to rescue her and killed the birdman. As Sedna and her father were escaping in his kayak the other birds found out what he had done and flew after them. Flapping their wings, the birds stirred up a great storm, and Sedna's father was terrified. He threw Sedna over the side of the kayak into the icy sea, to try to appease the angry birds. Sedna grabbed onto the side of the boat and tried to climb back in but her father cut off her fingers, which became the whales and seals and fish. Sedna now lives under the sea and rules over all the animals that live there. She has the body of a woman and the tail of a fish. If you look into the sea you might glimpse her, with her long hair like seaweed flowing this way and that in the current. The Inuit people venerate and fear her. Whenever they need more animals to eat, a shaman transforms himself into a fish and swims down to her. He combs the tangles from her hair and arranges it into fine braids. This makes her happy, and in return she releases animals from the ocean depths for the people to hunt.

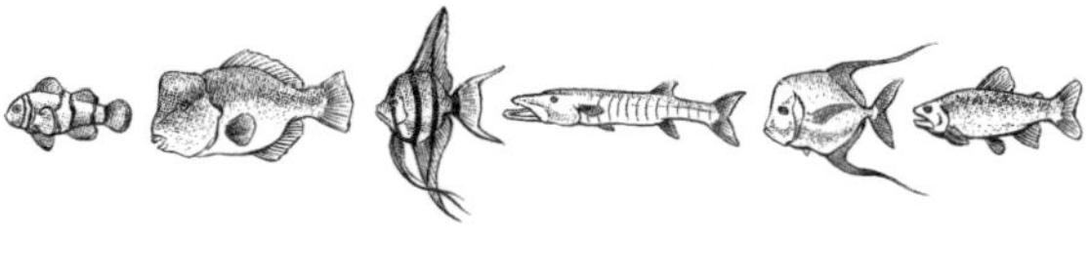

CHAPTER TWO:
A VIEW FROM THE DEEP – INTRODUCING THE FISH

CHAPTER TWO

A view from the deep – introducing the fish

A year after Charles Darwin returned from his voyage on HMS *Beagle*, in July 1837, he picked up a notebook and sketched a small stick-tree. He labelled the divided branches around the crown A, B, C and D, and next to it wrote the words 'I think'.

It would be another 22 years before he published *On the Origin of Species* and yet that twiggy sketch seems to have been Darwin's first attempt at visualising the pathway of evolution. The notion of a tree of life had been around for a long time, but Darwin was the first to propose a clear idea of how it all came about.

Fast forward to 2016, and a paper in the journal *Nature Microbiology* announced a 'New view of the tree of life'. It's based on gene sequences, and brings into play dozens of groups of bacteria and other microbes whose existence until recently went completely undetected. It's a far cry from Darwin's modest twigs, and doesn't look like a tree so much as a splayed frond of seaweed. Nevertheless, both trees are based on the same essential idea: that life gave rise to other life, and all living things are related.

When reading an evolutionary tree, the main points to remember are that the most ancient life forms are located at the bottom and give rise via natural selection to other lineages that split off and occupy their own branches. Also, the points where branches join the main trunk represent a common ancestor shared by the two lineages that split apart. On a human family tree, this is the equivalent of an uncle or a grandparent shared by cousins. In evolutionary

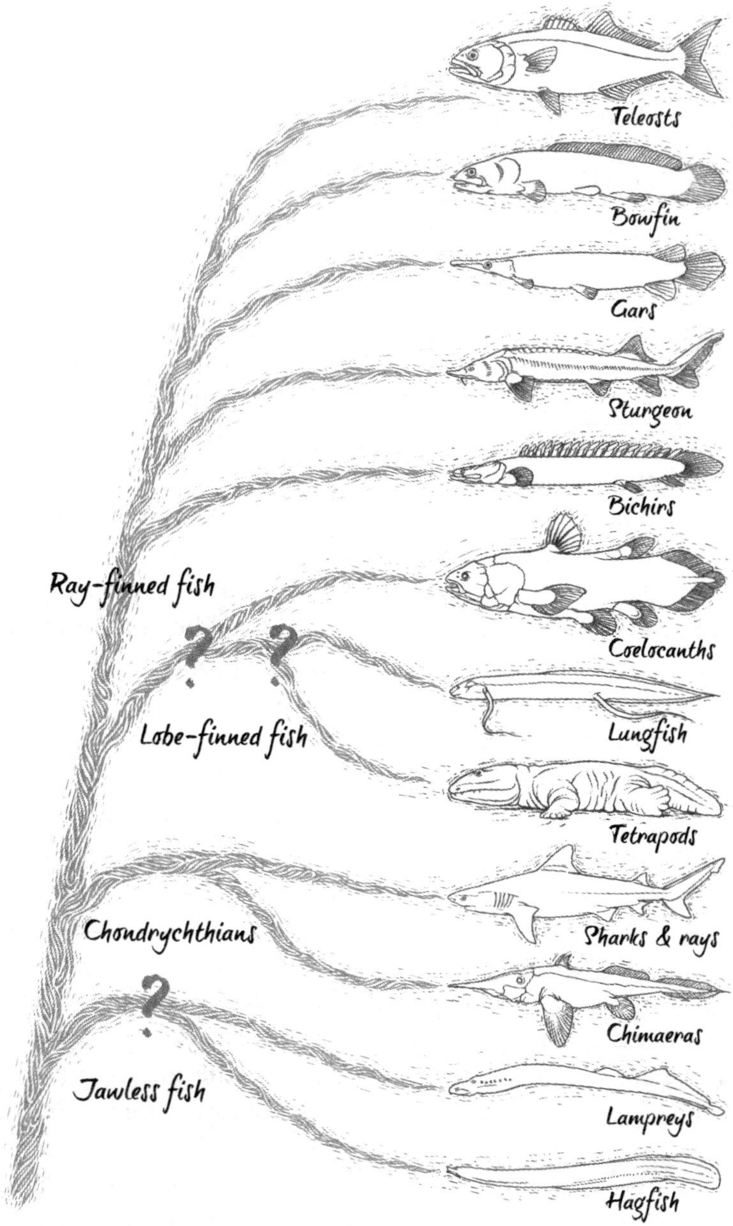

Fish evolutionary tree depicting likely relationships between 12 living groups.

studies, we rarely know what those ancestors looked like, or precisely when they existed. Nor indeed were the ancestors single organisms, but a population that split and gradually evolved into separate species.

If we zero in on part of the tree of life where fish are stationed, there's a thicket of branches. An ultimate version of the fish evolutionary tree would include all 30,000 species. Even the tree in Nelson's *Fishes of the World* spreads over several pages, depicting seven superclasses and 62 orders of fish. But fear not. To lead us through the immense diversity of fish life we don't need such a complex tree. We'll do just fine with a no-frills version with 12 branches and branchlets. That may seem like a drastic prune, but these 12 groups represent the essential divisions among the fish, and they'll make sure we don't miss anything significant.

We'll browse this tree not as if we were climbing an actual tree, starting from the ground and working our way upwards and outwards, towards the ends of dividing branches, but as if we've already reached the outermost boughs and are now clambering back towards *terra firma*. The species we'll meet in these 12 groups are by and large still alive today (evolutionary trees can include extinct members, too, but we'll get to those later) and, as we go, we'll pass junctions that represent increasingly ancient ancestors, shared by the branches above. In this way our exploration of the fish tree becomes a journey back through time. It means that as we move down the tree we'll encounter living fish groups with a direct lineage tracing back to ancestors that lived further and further in the past. And it means that we begin our journey with the latest fish group to split off from the others, one that also happens to be the most important today.

Think of a fish – any fish – and the chances are that swimming around in your mind is a teleost. This is the first group we encounter on the fish's evolutionary tree, and it's overwhelmingly the biggest. Teleosts account for roughly 96 per cent of all known fish species,* so it's no surprise that in this group we see the greatest diversity in the ways fish look, the things they do and the places they live.

Wherever there's water, there you can spot teleosts: Goldfish swim in garden ponds; gobies climb Hawaiian waterfalls, clinging to rocks with their lips; high in the Himalayan mountains are giant catfish that grow to the size of an adult human, and may occasionally eat one; silver shoals of anchovies, sardines and herring spiral through open seas, chased by master hunters, the swordfish, sailfish, tuna and Wahoo.†

There are teleosts living in glacial waters that swirl around Antarctica, the coldest ocean on the planet. Icefish survive in sub-zero temperatures and for years biologists guessed they must have some strategy to stop themselves freezing. The ocean itself doesn't freeze here because it's very salty, but living tissues can't tolerate such high salt levels. The fish had some other secret for staying supple. In the 1960s, Arthur DeVries, then at Stanford University in California, found molecules circulating in the icefish's bloodstream, called glycoproteins, which halt the growth of ice crystals. He had uncovered a fish version of antifreeze. Similar molecules have since been isolated from many other teleosts. Sculpins and flounders, herring and cod all have genes for making their own antifreezes, which they can switch on when the temperature drops.

---

*Teleosts have been shunted between various different taxonomic ranks; nowadays they're generally considered to lie somewhere between a sub-class and an order.

† Throughout this book, I capitalise formal common names, mainly to distinguish them from descriptions. So, there's a single species of Wahoo (*Acanthocybium solandri*) and many species of tuna.

Teleosts are also the deepest-living of all the fish. Down in the abyss, miles beneath the waves, are tripod spiderfish that perch on the seabed with three long fins and wait for prey to drift by. Further down still, cusk eels and snailfish vie to be not only the deepest fish but the deepest vertebrates of all. Exactly which group wins is frankly splitting hairs, since both have been spotted around eight kilometres (five miles) down, not quite at the bottom of the Marianas Trench, the oceans' deepest point. Both types of fish have soft, transparent skin and small, deep-set eyes that probably don't work especially well. Some cusk eels have such tiny eyes they've been named 'faceless fish'.

Even where there's very little water, teleosts still find a way to survive. In the middle of America's Death Valley, one of the hottest deserts on the planet, Devil's Hole Pupfish occupy a single underground lake, in a limestone cave. Since scientists started doing head counts in the 1970s, the number of these little blue fish has ranged from 35 to just over 500. At the latest count there were 115, giving the Devil's Hole Pupfish the dubious honour of being the rarest fish in the world. In April 2016, a group of drunken men broke into the pupfish's fenced-off cave, splashed about in the water, vomited and trod on several fish, killing at least one. To add insult to injury, they left behind a pair of underpants.

Teleosts can survive a lack of water by being mobile. When Walking Catfish from Southeast Asia find themselves trapped in a shrinking pond they simply trot off, hauling themselves overland on their fins, to find somewhere wetter. Mangrove killifish survive for months with no water at all, absorbing oxygen through their skin, and hiding in tree holes, empty crab burrows and coconut shells. When they get too hot, they leap into the air to cool down.

The biggest teleosts of all are sunfish, their bodies flattened discs up to three metres (10ft) across and weighing more than two tonnes (5,000lb).[*] They've been nicknamed 'swimming heads'; they have no tails, and swim by flicking a tall dorsal and anal fin from side to side. Their common name, sunfish, comes from their habit of sunbathing on their sides at the sea surface. This added to their reputation as sedate, loafing animals, but in 2015 Itsumi Nakamura and colleagues at the University of Tokyo fixed temperature probes, accelerometers and little cameras to sunfish and discovered they make energetic dives into deep, cold waters chasing jellyfish, then they lie in the sun afterwards to warm up.

The name *teleost* refers to the fact that they perch on this outermost branch of the fish evolutionary tree. They're the ones with 'perfect bones' (from Ancient Greek words *teleios* meaning perfect or full grown, and *osteon* meaning bone), although there's nothing inherently perfect about them; they just happen to occupy the branch that split off most recently.

All of these various fish are collected together because they share a set of characteristics. Their 'perfect' bones are less dense than their ancestors', with internal cross-struts that make them strong but light. Their backbone runs all along the body and comes to an end at a region just before the tail called the caudal peduncle. Here the arrangement of bones is such that their tails are stiffened compared to their more distant ancestors.[†] This generally makes teleosts

[*] Until recently, it was thought there were four sunfish species, the best known being the Ocean Sunfish (*Mola mola*). In 2017, a fifth species was tracked down after a four-year hunt by Marianne Nyegaard at Murdoch University in Australia. This one is *Mola tecta* (from the Latin for 'hidden'), the Hoodwinker Sunfish.

[†] This is why, should you want to slap someone with a dead fish, I would recommend you use a teleost.

good swimmers; they propel themselves with powerful beats of their tail without needing to swing their whole body from side to side.

Teleosts are generally covered in ultra-thin, lightweight scales, comprised of a composite of collagen and hydroxyapatite, the calcium-based mineral that makes up much of human bones. Scales grow in overlapping rows from head to tail, like roof tiles, giving teleosts tough, flexible suits of armour.

There's also a certain arrangement of bones in teleost jaws that allows them to fling their mouths forwards to suck in prey. This has helped them diversify their diet and means they can eat just about anything; they'll nibble on plankton, graze on seaweed, eat each other (dead or alive), munch leaves, slurp dirt and chew seeds. The rest of the fish we'll meet are much fussier – most of them are strict carnivores.

Within the teleosts are dozens of sub-groups. There are some 1,600 species of piranhas, tetras, characins and pencilfish, more than 3,000 carp, minnows and loaches, and a similar number of catfish, including mountain, velvet and bumblebee catfish. There are over 550 cod, 700 flatfish and another 800 eels.

Two of the biggest teleost sub-groups bring together many of the cast of animals that feature in this book. One rambling group (the Percoidea) is home to ponyfish that glow in the dark, archerfish that shoot water, leaf-fish that pretend not to be fish and croakers that make a racket. Many coral-reef fish are here too: groupers, butterflyfish, angelfish, goatfish, snappers and dottybacks.

Wrasse and damselfish, parrotfish and rabbitfish – named respectively for their beaky and bucktoothed dentitions – join together in another large group (the Labroidei) along with cichlids, which inhabit freshwaters worldwide but most famously as a flock of species in the Great Lakes of Africa. Some time between two and 10 million years ago a new split opened up in the Earth's crust, along Africa's Rift

Valley, and filled with water to form Lakes Malawi, Victoria and Tanganyika. Cichlids were among the first fish to move in, and subsequently they've evolved into some 1,700 endemic species found nowhere else on the planet. Some of them evolved fast and furiously. It's thought Lake Victoria was completely dry 12,500 years ago, which means around 500 endemic species must have arisen since then.

With so many perfect-boned fish, it's perhaps easier to point out which animals you might spot that are something other than teleosts. To meet them we need to move on to the next branch down the fish evolutionary tree.

Take a look in a quiet backwater of a river in the eastern United States, perhaps the St Lawrence or the Mississippi, and you might catch sight of a lone evolutionary survivor. You'll probably need to peer closely among plant roots or under logs to find one. It is cylindrical in shape, dappled olive brown in colour and usually around 50cm (20in) long when fully grown. If there's a spot on its tail, with an orange ring around it, then you have a young male. It can swim forwards and backwards with equal ease, undulating the dorsal fin that runs all along its back.

This is a Bowfin, the teleosts' closet living relative. Around 200 million years ago, in the late Triassic, bowfins shared a common ancestor with the teleosts. They went on to flourish in marine and freshwaters from America to Europe, Asia to Africa. Now, though, you'll only see a single Bowfin species in the margins of rivers, lakes and swamps in North America.

A mix of body parts places the Bowfin in its own separate group (the Amiiformes). They share some characteristics with teleosts, such as gills for extracting oxygen from water; a gas-filled balloon known as the swim bladder, which acts like an internal flotation device; and a series of fluid-filled pores and canals called the lateral line, which detect movements and vibrations in the water (more on those later). Bowfins also have a handful of other, more unusual features, including the way they use their swim bladder not just to float around but to breathe dry air. This is more reminiscent of fish further down the evolutionary tree, including the gars, which lie on the next branch down.

In the shallow, weedy rivers and lakes of North America, as well as Bowfins, you might spot a member of another group of fish survivors that once lived worldwide. There are seven species of gar, which share a common ancestor with Bowfins that lived roughly 260 million years ago. Gars are covered in interlocking, glassy scales, which Native Americans traditionally used to cover breastplates and to make tips for their arrows. The most notorious gar, and the biggest, is the Alligator Gar. With a long snout and two nostrils at the end, this two-metre (6.5ft) fish looks distinctly reptilian. The double rows of teeth inside its mouth are an instant giveaway that it's not a real alligator, but from a distance it can still have people easily fooled. In 2010, dozens of Alligator Gars were spotted swimming in a lake in Hong Kong, a long way from their original American homes. Locals panicked, assuming they were true crocodilians, and officials quickly came to take them away. It's assumed the wandering gars had been set free by aquarium-keepers who weren't prepared for the enormous size their pets would reach.

Eggs are the most celebrated parts of the fish we encounter next on our journey down the fish evolutionary tree. On the fourth branch down are 27 living species of sturgeon. Most of them haven't changed a whole lot in appearance since their ancestors shared Jurassic seas with ichthyosaurs and plesiosaurs, and maybe sensed the footfall of dinosaurs tromping the edges of brackish lagoons. Instead of scales sturgeons have rows of spiky plates, called scutes; they have fleshy, upturned snouts, speckled in electro-sensitive pores, with four whiskers hanging down.

Sturgeon are northern-hemisphere residents. If you're lucky you might spot them in the same American waters as Bowfin and gars; they also live across Europe and Asia, in rivers and lakes from the Atlantic to the Pacific. Throughout their range, sturgeons haven't been doing so well, ever since people developed a taste for their eggs, known as caviar. Favoured species include Kalugas from the Amur River, which rises in the mountains of northeast China and flows to the Sea of Okhotsk in the north Pacific. Sevruga Sturgeons, from rivers draining into the Caspian, Azov and Black Seas in eastern Europe and Asia, are also prized for their eggs. In the same waters swim Beluga Sturgeon, whose eggs are considered the finest caviar, commanding the highest prices. An individual female can grow up to eight metres (26ft) long, bigger than her mammalian namesake the Beluga Whale; her ovaries can account for a quarter of her body weight, and she'll lay millions of eggs in a single clutch. A record-breaking Beluga was killed in Russia in 1924, weighing more than 1.2 tonnes and with 245kg (540lb) caviar inside her; try buying that much caviar on the open market today and it would set you back at least a million pounds, perhaps two.

The demand for caviar is one reason why, in 2010, sturgeon were declared the most threatened group of animals in the world, with 23 out of 27 species at risk of extinction.* Syr Darya Shovelnose Sturgeons, with long whip-like tails, haven't been seen alive since the 1960s. Beluga Sturgeon are considered to be critically endangered.

---

*This is according to the International Union for the Conservation of Nature (IUCN) Red List of Threatened Species. Most caviar-hunters don't wait for the females to lay their eggs, but cut them out. There are a few initiatives to produce sustainable caviar.

The only species you stand a good chance of encountering is the White Sturgeon from the US Pacific coast. At up to six metres (20ft) long, they're the biggest freshwater fish in North America, although they more usually reach only half that size and actually spend more of their time at sea, close to shore, before swimming inland to spawn.

Dams blocking the sturgeons' paths as they try to migrate to their spawning grounds add to their modern-day troubles, as do the increasingly polluted waters they swim through. And in general, sturgeon are poorly equipped for dealing with any of these problems. They can take decades to reach maturity, and even then females only spawn once every five years or so. Compared to faster-growing species, sturgeon populations don't recover quickly. On the plus side, sturgeon can live for a century or more, which means that if a few remaining individuals are wandering around in the wild, there's a chance they might eventually find each other and successfully spawn.

In even worse shape are two close relatives of sturgeons, also found on this same evolutionary branch (the Acipenseriformes). One lives in America, the other in China – or at least it did. Scientists fear that the Chinese Paddlefish is now extinct. It was last seen alive in 2003. That's despite a team of biologists spending three years, up to 2009, looking for it along the Yangtze River, from high in the Tibetan plateau all the way to Shanghai. All they came back with were two sonar readings that detected lumps in the water of around the right size that could, maybe, have been paddlefish.

American Paddlefish, also known as Spoonbill Catfish, are doing a little better across their range in the lakes and braided channels of the Mississippi basin. A third of their two-metre (6.5ft) long body consists of a broad, flattened snout (which looks a bit like a paddle, hence their name), supported by a lacy network of star-shaped bones that lie beneath their scaleless skin. It was only recently that scientists worked out the purpose of this odd protuberance.

The paddle is covered in dimples equipped with sensory receptors that detect weak electric fields. As a paddlefish sweeps its snout through the water, it homes in on pulses from wriggling water fleas close by and swoops in, its mouth wide open like a trapdoor.

Efforts are underway in North America to restore paddlefish to their former glory. They used to have a much wider range, throughout the Great Lakes and in at least four states where they're no longer found. Many reservoirs are stocked with farm-reared paddlefish, partly so anglers can catch them, but these are located in spots where there's nowhere for these fish to spawn. For that they need fast-flowing water and clean gravel. Many of America's paddlefish are just old fish, getting older.

Search for a missing link

Taking a step further down the fish evolutionary tree, we meet the first in a series of fish that for a long time had taxonomists baffled. Bichirs* look like small snakes with smiles. There are 12 species, known as dragonfish in the pet trade, and to see one in the wild you'll have to go and look in rivers and swamps in Africa. They're covered in shiny scales and have a long dorsal fin snipped into sections, known as finlets; they swim by fluttering their wide, fan-like pectoral fins. Bichirs mainly breathe air through a pair of lungs, inhaling through their mouth and exhaling through holes called spiracles on the top of their head. They have rudimentary gills but in stagnant water, if they don't have access to the surface to breathe air, they drown.

When they were first discovered in the River Nile in 1802, anatomists had never seen such an odd mixture of features and they sparked an important question: are bichirs a link between fish and amphibians?

The lack of intermediate stages between different animal groups was something Charles Darwin contemplated at

---

* Pronounced 'bi-shears'.

great length. Finding these apparently 'missing links', whether in fossils or living creatures, would support his theory of species giving rise to other species, and help map out the shape of the tree of life. Of particular interest were the links between vertebrates that live in water and those on land, the tetrapods, and the question of how exactly our ancestors adapted to a terrestrial way of life.

Towards the end of the 19th century, it became popular to study animal embryos to gain clues about the pathways of evolution. The idea was that animal groups look distinct at a microscopic level during the first steps of life, as fertilised eggs divide into a bundles of cells. To work out whether bichirs were fish or amphibians, or somewhere in between the two, scientists were convinced they needed a bichir embryo – something that wasn't easy to come by. These animals live in places like the Congo and Nile basins, which can still be difficult and dangerous to get to today. But that didn't put off two men who, more than a century ago, were determined to fill in this zoological gap.

One was an Englishman, John Budgett. As a young boy, he was already a keen zoologist. He kept all sorts of pets at home and built a small museum, filled with stuffed animals and skeletons he prepared himself, including a cow, a deer and his family's Shetland pony. He frequently visited his local zoo to check on the health of any sick animals, hoping for new specimens.

In 1894, Budgett went to Cambridge University to study zoology, but he was soon lured further afield. His first taste of exploration came in 1896 when he accompanied John Graham Kerr, another Cambridge student, on a year-long expedition to Paraguay to collect lungfish (a fish group which we'll meet shortly), in swampy, insect-infested conditions that Budgett would become familiar with. Kerr and Budgett found their first lungfish without effort, when locals served some up for their first dinner. Kerr later wrote that the lungfish had been 'most tasty'. Returning from Paraguay, Budgett scraped through his final year exams in

Cambridge and made plans for his own expedition, this time to find bichirs.

Meanwhile, apparently unbeknown to Budgett, another bichir hunter set out to find embryos of this possible missing link. In 1898, Nathan Harrington from Columbia University in New York spent four months in Egypt, scouring the Nile for bichirs. He found mature adults and tried many times to artificially fertilise their eggs – taking eggs from a female and bathing them in a male's sperm – but without success. On a return trip to Egypt, Harrington was overcome by fever and in 1899, aged 29, he died.

John Budgett had thought of going to the Nile but on a friend's recommendation went instead in October 1898 to the other side of the African continent, to the small nation of The Gambia, at the time a British colony. He spent eight months far inland along the River Gambia, much of it in pouring rain, undertaking the same fruitless research as Harrington. At least Budgett learned much about catching these unusual, nocturnal animals, and he pinpointed their breeding season. Now he knew exactly when to return to Africa.

At the end of that first trip, Budgett brought two living bichirs back to England and they lived on for another three years, carefully tended by his brother, Herbert. The captive bichirs put on courtship displays, but never produced any young.

In 1900, undeterred by repeated bouts of malaria picked up during his travels, Budgett went back to The Gambia, this time at the height of the rainy season in June when he was sure the bichirs should be mating. Yet again, despite three months of searching, he didn't find any embryos. He tried again, in 1902, this time in eastern Africa, in Uganda and Kenya, but once again he came home empty handed.

Then, the following year, Budgett's luck finally began to change. He sailed back to West Africa and travelled by paddle steamer up the River Niger. The going was hard. 'It rains almost continuously, everything is mildew and rust,' Budgett wrote in his diary. 'The depression of this vapour-bath is almost unbearable.'

Finally, after four gruelling expeditions, Budgett found what he had been searching for, but it came at great cost. In Nigeria, on 26 August 1903, he successfully fertilised bichir eggs and watched down his microscope as the transparent spheres began to split and cleave into living balls of cells. Two days later he sent a letter to his old friend, Graham Kerr, telling him that the resulting embryos were 'astoundingly frog-like'. They were complete (meaning the whole eggs divides and not just part of it) and equal (the egg splits into cells of equal size) and the bundle began to fold in a way seen in frog embryos. But by the time Budgett prepared to head home with his prized preserved embryos, he was again suffering from malaria.

Back in Cambridge, on 9 January 1904, having just finished a series of intricate bichir embryo drawings, Budgett showed the first signs of blackwater fever, a deadly complication of malaria in which red blood cells burst in the bloodstream. Ten days later, on the day he was due to present his findings to the Zoological Society of London, John Budgett died.

A collection of preserved eggs and embryos and some detailed drawings were all that Budgett left behind to mark his ultimately fatal quest. There was no manuscript or notes. Four years later it was another scientist, Edwin Stephen Goodrich from Oxford University, who gathered together everything then known about bichirs and announced that they are not the direct ancestors of frogs – just very strange fish. Many of their peculiar features evolved separately along their own branch of the fish evolutionary tree, including their frog-like early development and their ability to regrow a severed limb.

Much later, in 1996, DNA sequencing confirmed that bichirs are not a missing link between fish and frogs, but they are the earliest division of the ray-finned fish, fish that have fins made up of spines sticking up from the base with skin stretched between them.[*] Interest in bichirs and their embryos faded but ichthyologists remained preoccupied for much longer with another enigmatic group that is situated on the next branch down the fish evolutionary tree.

Lungfish have long been mistaken for other things. A fossilised lungfish tooth was described in 1811 as a piece of tortoise shell. In 1833, Swiss scientist Louis Agassiz, probably the greatest ever authority on ancient fish, identified another lungfish fossil as a type of shark (he later changed his mind). When the first living lungfish showed up in 1836, at the mouth of the Amazon River, experts back in Europe thought it was a reptile, due to the snippet of lung tissue still hanging to the gutted specimen. The following year a different variety was discovered in Africa and, based on the structure of its heart, it was declared amphibian.

For another three decades, the lungfish debate rumbled on, with experts picking and choosing sides. Richard Owen, famous for founding London's Natural History Museum and coining the word 'dinosaur', was utterly convinced of the fishy nature of lungfish, pushing them away from reptiles. 'Not by its gills, not by its air bladders … nor its extremities nor its skin nor its eyes nor its ears,' he wrote, 'but simply by its nose.' He was sure that a reptile's nose has two openings, while a fish merely has a blind-ended sac, something he thought he saw in lungfish.[†]

[*] Also known as actinopterygians, a broader grouping including all the fish we've met so far, higher up on the evolutionary tree.

[†] Lungfish in fact have paired nostrils that connect through to the mouth and draw water over a layer of skin with receptors that bind to odour molecules in the water. Most teleost noses are blind-ended sacs, with two nostrils to draw water in and out again.

Today all six known lungfish species are confined to slow-moving rivers, swamps and freshwater pools in Africa, South America and Australia. They all have elongated, eely bodies, some up to two metres (6.5ft) long, and some have pelvic and pectoral fins like strings of spaghetti. Only the Australian Lungfish has gills that still work, while all the others rely entirely on their paired lungs for oxygen. So, like bichirs, it's quite possible for these fish to drown. It does mean, however, that they can survive without water. In Africa and South America, lungfish can tough out long dry seasons by chewing themselves a burrow in the mud, filling it with mucus, and curling up inside; they can survive this way for up to four years. When finally the rains return, the lungfish emerge from the mud and eat whatever they come across, often another sleepy lungfish who has also just woken up. And they can live for a long time – at least in captivity. A lungfish called Granddad was taken from the wild in Australia in 1933 and kept in the Chicago Aquarium, where he died in 2017.

For a while it was thought that there might be a seventh lungfish. A few years after the first Australian species was found, another came to light in northern Queensland in 1872 when Karl Staiger, director of the Brisbane Museum, ate one for breakfast. The lungfish was 45cm (17in) long, covered in large scales, and had a flattened snout, curiously similar to that of the platypus. Staiger paused, fork in hand, just long enough for someone to sketch the strange fish and write a few notes, before he tucked in. The notes and drawing went to French naturalist Francis de Castelnau, who named the new species *Ompax spatuloides* and likened it to the Alligator Gar of North America but decided it was a new type of lungfish. A second specimen was never found, but the strange creature did emerge one more time, in a letter sent to a Sydney newspaper almost 60 years later. The truth was revealed that Staiger's meal had in fact been a hoax, stitched together from a mullet's body, an eel's tail, a platypus's bill and the head of an Australian lungfish.

Up to the present day, lungfish have remained caught in the cross-winds of research trends, batted from place to place in fossil and evolutionary studies, embryology, molecular sequencing and more. Still there are matters undecided. One question lungfish pose is whether fish first evolved lungs or swim bladders. Did they evolve lungs first, organs rich with blood vessels and permeable to gases, then later co-opt them as airtight flotation devices? Are lungs modified swim bladders, or did the two organs arise independently?

In embryos, swim bladders and lungs both develop from a pocket in the gut. No fish have both organs so, like Clark Kent and Superman, it seems likely that one is a version of the other.

You may already be thinking of swim bladders as a distinctly fishy characteristic, a feature of all the fish we've met so far. However, lungfish don't have them, suggesting that lungs could in fact have been around for the longest.

Research from a few years ago backed up this idea, when Sarah Longo from Cornell University in New York took a lungfish, a paddlefish, a sturgeon, a Bowfin, a gar and a bichir and put them one by one inside a CT scanner.* This let her scrutinise the detailed arrangement of their blood vessels, revealing key similarities between the fish with lungs (the lungfish, Bowfin and bichir) and those with swim bladders (the sturgeon, gar and paddlefish).† Longo found that all these organs are hooked up to a pair of pulmonary arteries, the same vessel that transports blood from your heart to your lungs. In sturgeon and gars these vessels are vestigial and hadn't been detected before. Discovering that their swim bladders are in fact connected

---

* These machines are used in medical imaging to produce high-resolution slices through intact living tissue; these can be combined into 3D images.

† The Bowfins' 'lungs' are in fact thought to be a modified swim bladder.

to the same blood vessels as lungs in more ancient fish provides another strand of evidence that lungs did indeed evolve first and were later adapted to form the swim bladder.

Lungfish share this branch of the fish evolutionary tree with two close living relatives. This trio are together known as lobe-finned fish (or sarcopterygians) and they differ from ray-finned fish chiefly in that they have fleshy fins that join via a bony connection to the backbone at the hip and shoulder. As we'll see, it's not exactly clear which group split off on the first twig along this branch, but all of them are without doubt important in the history of fish.

Coelacanths* are perhaps most famous today as fish that were thought to have been extinct for millions of years. That was until 1938, when a South African biologist, Marjorie Courtenay-Latimer, was paying one of her regular visits to the local docks to see what fishermen had trawled up, when she spotted something strange. It was a huge, mauve fish with iridescent silver markings, a tail with three lobes and four large, fleshy fins. She carted off the two-metre (6.5ft) fish in a wheelbarrow to find somewhere to get its remains preserved. Her finding was as unexpected and dramatic as a living *Velociraptor* ambling out of a far-flung desert. Eventually, South African ichthyologist J. L. B. Smith created an entire new genus for the fish and named it after her – *Latimeria.*

It's well established now that at least two coelacanth species inhabit deep, sunken volcanic slopes pocked with caves. Marjorie's species lives around the Comoros Islands, off the coasts of Madagascar, Mozambique and South Africa, while a second species was spotted in 1998 by another biologist, Mark Erdman, in another fish market,

* Pronounced 'sea-la-canths'.

this time in Indonesia. These are the only known descendants of at least 80 coelacanth species that once roamed oceans and freshwaters around the world.*

Since their rediscovery, more details of coelacanth's lives have come to light. Females produce enormous eggs, bigger than baseballs, which hatch and develop inside them for up to three years before being born (a process known as ovoviviparity). Adult coelacanths spend their days huddled together in caves down at 250m (820 ft); it's no wonder they stayed hidden from scientists for so long. At night, they venture deeper, to 500m (1,640ft), to hunt for fish and squid. Initially it was thought they might stalk the seabed on their fleshy fins, but film footage shot from mini submarines has shown them drifting along, sculling with their four paired fins in a diagonal pattern, the same way a lizard moves its legs. But coelacanths were not the direct ancestors of amphibians, or reptiles, or anything else with four legs. For that we need to look at another group of lobe-finned fish and another of the lungfish's close relatives.

Back in the Devonian, around 380 million years ago, coelacanths and lungfish shared the seas with another group of lobe-finned fish, a renegade troupe called the tetrapodomorphs.† Some looked similar to lungfish and paddled around open water. Others looked more like huge salamanders; in place of fins they had legs and arms, hands and feet. They clambered about boggy, marshy plants and could have lifted up their heads, looked over their shoulders and waved eight little fingers.

---

* There's a risk that coelacanths could yet go extinct after all. The two species are listed by the International Union for the Conservation of Nature as Critically Endangered and Vulnerable to extinction, mainly from incidental capture in small-scale fisheries. There's also been growing demand from idiots who think that eating coelacanths will make them live for longer.

† The tetrapodomorphs are now all extinct; they included such beasts as the *Eusthenopteron*, *Panderichthys*, *Ichthyostega* and *Acanthostega*.

All these animals, now long gone, began to show themselves to palaeontologists through a series of remarkable fossil finds. The most recent was *Tiktaalik*, found in 2004 in Ellesmere Island in far northern Canada. It looked like a cross between a lungfish and a small crocodile and probably hung about in shallow waters, dashing out to avoid the jaws of huge predatory fish and to snap up insects that were beginning to crawl and scuttle about on land.

Tiktaalik and its Devonian siblings present an elegant sequence of creatures that shifted from a water-based lifestyle to a land-based one. Charles Darwin would have had a field day with them. The arrangement of their bones shows that tetrapod ancestors didn't evolve all at once but gradually, in stages, as these lobe-finned fish became increasingly adapted to life at watery margins and beyond. Here is the fish-frog, water-land transition that palaeontologists had been searching for.

It's not known exactly how these transitional fish clambered out of water all those years ago, but recent studies of living fish are providing enticing new clues, including from bichirs, those strange fish that John Budgett devoted his life to finding. While bichirs don't represent the direct ancestors of tetrapods, they are helping to reveal how extinct fish may have learned to walk.

When water levels recede, bichirs scramble around using their pectoral fins. In 2014, Emily Standen at McGill University in Montreal discovered they can quite quickly improve their walking abilities. She reared some bichirs in a regular, water-filled aquarium tank, and others in a tank with only a thin film of water not deep enough to swim in. After a year, the ones from the drier tank had modified their walking movements compared to the swimmers; they lifted their heads higher, planted their fins more firmly and slipped less often. Not only that, but their bones and muscles changed as they adapted to a perambulatory lifestyle. Similar anatomical reshaping shows up in the fossilised bones of *Tiktaalik* and its close relatives as they adjusted to life on land. This plasticity in the way fish move hasn't been demonstrated before; it shows just how flexible they can be, and how fast they can adapt to a changing world.

Whether coelacanths or lungfish are the closest living relatives to tetrapods remains a matter that's firmly undecided. For decades, taxonomists have been constantly reshuffling these twigs on the tree of life and the wrangling continues today, even with the latest genetic studies. In 2013 the coelacanth genome was sequenced, providing all the pieces of the puzzle, but still the picture remains unclear. Part of the problem lies in the choice of another animal group that's used for comparison, the so-called outgroup. Using elasmobranchs (sharks and rays) as the outgroup, two separate research teams pushed lungfish into prime position as the sister group to tetrapods. That all seemed well and good until 2016, when another analysis switched things around. A team from Japan swapped in teleosts as the outgroup and consequently put coelacanths back next to the tetrapods. However, in 2017 the same Japanese team had another go, using gars and Bowfins as the outgroup, and this time reinstated the lungfish-tetrapod alliance. This could change once again but, as it stands, those mud-chewing, air-breathing lungfish are the human's nearest living fish relatives. The ancestors we share with them lived roughly 400 million years ago.

Picking up where we left off and continuing on down the fish evolutionary tree, we reach branch nine out of 12, which split off from the tree at least 450 million years ago. This is the only fish group besides teleosts that contains more than a smattering of living species, and so far they've all been waiting patiently in the wings. It's time to formally welcome the elasmobranchs. The name comes from Greek words meaning 'beaten metal gills', but perhaps in the sense of beaten metal being somewhat bendy and elastic, a nod to this group's soft, cartilaginous skeletons.

There are around a thousand species of elasmobranchs to spot, as they cruise through and sit in the oceans today. Roughly half of them are sharks, the rounded, upright ones with gills on the sides of their bodies. There are familiar sharks, like Great Whites, Salmon Sharks and makos (all types of mackerel shark), Blue Sharks, Oceanic Whitetips and various reef sharks (all requiem sharks). And there are plenty of obscure and little-known species: there are Nervous Sharks, Graceful Sharks, Shy Sharks and Blind Sharks (so-called not because they're blind, but because they shut their tiny eyes when brought up from the depths into bright light); there are Zebra Sharks and Crocodile Sharks, Grinning and Crying Catsharks, Cow Sharks and Frog Sharks.

The other half of the elasmobranch group are the rays and skates, all of them flattened top to bottom, with gills and mouth on their underside and spiracles on the top to breathe through. There are stingrays, maskrays, electric rays and numbfish, Sapphire Skates and Munchkin Skates, Fanskates and Slime Skates. Some are kite-shaped, others form a perfect circle. Many lie flat on the sea- or riverbed,* covering most of themselves in sediment except for a pair of eyes sticking up. Some swim through open water, like devil rays and eagle rays, flapping wide pectoral fins like wings.

Some sharks are flat and lie on their bellies, like rays. Wobbegongs have dappled skin and mossy beards that blend into reefs, where they sit and wait for prey to wander by. Sawsharks have long snouts fringed in teeth, which they use to root for prey hidden in the seabed. They look a lot like sawfish, but you can tell them apart by locating

* Unlike sharks, which almost all live in the sea, there are lots of freshwater stingrays.

their gills: sawfish are rays with gills underneath; sawsharks are sharks with gills on the side.

Ever since I started scuba-diving I'd been dying to see a shark. The idea of them never scared me. There's the thrill of seeing large, wild animals which I rarely come across at home in Britain, except for the occasional deer. I was also determined to prove people wrong, including a few friends and family, who have been led to believe that all sharks are dangerous beasts with voracious appetites for human flesh.

When I went to Belize on a two-month diving expedition I was sure I would finally spot a shark. But after diving two or three times every day on the offshore reefs, I was beginning to lose hope. Then, just a few days before I left, I was finally in the right place at the right time. The wait had made it all the more satisfying, but also at the back of my mind I knew this uncommon encounter was a sure sign of decades of overfishing.

My first shark experience took place while I was on a drift dive, riding a fast current and flying along much faster than normal swimming speed. Ahead I spotted a huge stingray and next to it, snoozing peacefully on the sand, was a Nurse Shark (proof that at least some sharks don't suffocate when they stop swimming). As I drifted past I had just long enough to see that the shark was bigger than me, it had smooth grey skin, little eyes and a trim moustache dangling from its blunt-ended snout. It lifted its head and, with slow beats of its tail, swam into the current and out of sight.

Nurse Sharks are mostly nocturnal, spending the night rummaging around reefs, hunting for crabs and molluscs hiding in the seabed. Like all elasmobranchs, Nurse Sharks have sensory pores that can detect the weak electric fields generated by living bodies. When they're not hunting, Nurse Sharks spend a lot of time just sitting. Their inactivity minimises their energy requirements and they can survive when there's not much food around. Sharks have very low metabolic rates compared to teleosts; they use up less oxygen, burn less fuel and get away with eating a lot less. After a Great White Shark has chewed on the floating carcass of a dead whale it probably won't have to eat again for another six weeks. A Salmon needs to eat at least four times more than a similar-sized shark. And Nurse Sharks are among the most energy efficient of all the sharks, with the lowest metabolic rate measured so far. A 2016 study revealed that they use about 80 per cent less oxygen per hour, per kilo of body weight, than a fast-paced shark like a Mako.

This bid to save energy is a theme that runs throughout elasmobranch biology, and it's a key to their great success. They've lightened their skeletons, replacing heavy bone with lightweight cartilage, the same rubbery tissue that your nose and ears are made of. Sharks also boost their swimming efficiency with their enormous, oily livers that slow their rate of sinking and help keep them afloat, in a similar way to the teleosts' swim bladders. A Basking Shark that weighs a tonne in dry air only weighs 3.3kg (7lb) when it's underwater because of its buoyant liver. In the 20th century, Basking Sharks were hunted for their liver oil, as a source of vitamin A and as a high-grade lubricant for the aviation industry.* Even today, various species of deep sea sharks are targeted for their oil, which is rich in squalene, a molecule that's used to make cosmetics and haemorrhoid cream.†

Elasmobranchs gain further efficiency from their skin. Rather than teleost scales, they're covered in tiny, sculpted denticles – highly modified teeth – that reduce drag and

* Basking Sharks are now protected under EU law.

† Squalene is also sold by health companies in capsule form, despite scanty evidence of any known health benefits.

help them slip through the water. This makes them not only more streamlined but also more silent, so they can sneak up on their prey.

The super-efficient lives of elasmobranchs can go on for a long time. Sawfish live for 40 years, dogfish for a century, and in 2016, Greenland Sharks were recognised as the longest-lived vertebrates. These fish live in deep Arctic waters where, at up to seven metres (23ft) long with mottled grey skin and only a very small dorsal fin, they look more like giant seals than sharks. And they've revealed their immense age in their eyes. Atmospheric tests of thermonuclear weapons in the 1960s sprayed a bomb pulse into the oceans, which has since trickled through marine ecosystems. This radioactive time stamp laid down in the lenses inside the Greenland Sharks' eyes helped researchers work out that they can live for at least 270, and perhaps closer to 400 years, given a chance. And in these long, slow lives, many sharks take their time to get going. Great Whites only become sexually mature when they're teenagers, and Greenland Sharks may mate for the first time when they're 150 years old.

When they finally get around to making more of themselves, elasmobranchs usually meet up, form couples and mate, something that most other fish, such as teleosts, don't do.* Occasionally, divers find themselves in the right place at the right time to watch this happening, and catch glimpses of courtship rituals. Reef Manta Rays parade around, with a receptive female at the front being trailed by dozens of eager males. She will twist and turn and even leap from the sea, perhaps testing her potential mates to see which is the strongest. Eventually, she chooses a mate and lets him slide his mouth onto one of her pectoral fins; he bites down with his tiny teeth (which he doesn't use for

* With a few exceptions, the usual way of things for fish is for a female to lay eggs in the water and a male to add a squirt of sperm to externally fertilise them.

eating). Another male might come along and try to knock him off, but if he keeps a firm grip he'll flip his body underneath and hold his belly close to hers.

Like all male elasmobranchs, mantas have a pair of modified pelvic fins, known as claspers, that dangle underneath. They look like stretched-out testicles but they act like a penis, transferring sperm into the female's body. Like coelacanths, manta rays are ovoviviparous, with the fertilised eggs hatching inside the female and staying there. After a year of gestation, the fully-formed manta pups are born, either as singletons or occasionally twins, wrapped up in their wide fins like a baby in a blanket.*

Hammerhead sharks, Blue Sharks and various others are viviparous, meaning the females provide unborn young with food and oxygen from a placenta via an umbilical cord, in a similar way to mammals. They too give birth to a small number of fully-formed young. A third breeding option for elasmobranchs is to lay egg cases on the seabed (rather than ovoviviparity, this is known as oviparity – egg-laying). Also known as 'mermaid's purses', the leathery egg cases look like giant ravioli pasta shapes, and they often wash up empty on beaches after the pups have climbed out. From the shape and size of the egg case you can work out which species it came from. Catshark egg cases have long curling tendrils at each end, which anchor them to strands of seaweed. Around the coasts of Australia, Port Jackson Sharks lay spiral-shaped egg cases, then pick them up in their mouths and wedge them between rocks. Ten months later, the pups hatch at around 20cm (7in), big enough to reach from top to bottom on this page.

For a handful of elasmobranchs there's a fourth option: to go it alone. Female Bonnethead Sharks, Zebra Sharks, Swellsharks, bamboo sharks and sawfish are all known to have given birth with no male contribution. Their

---

* The name 'manta' comes from the Spanish word for blanket.

unfertilised eggs have, occasionally, developed directly into embryos, giving rise to genetically identical offspring, a useful tactic to switch to when mates are hard to come by. Insects often employ this trick, as do a few reptiles, birds and amphibians (as far as we know, mammals can't to this without a lot of help from modern cloning techniques).

No matter how they're born, an important aspect in the lives of elasmobranchs – which may seem rather obvious – is being big. Most stingrays are at least the size of a dustbin lid and they can be much bigger. In 2015, a Freshwater Whipray was captured in a river in Thailand that was more than 2.4m (7.9ft) in diameter and 4m (13ft) from nose to tail. It looked like an elephant that had been melted down, scooped up into a ball and then trodden on. The sting at the end of its tail was 38cm (15in) long, loaded with venom, and made from a hugely enlarged and modified denticle, the tooth-like structure normally seen in elasmobranch skin. Contrary to popular belief, stingrays only use their stings in defence, not attack.

As for the sharks, Dwarf Lantern Sharks are small enough that you could easily put one in your pocket, but most fully-grown sharks are much bigger. They include the three biggest fish alive today, the Whale, Basking and Megamouth Sharks (which range between roughly seven and twenty metres, or 22–65ft). Meanwhile, more than half of all newborn sharks arrive in the sea longer than the average human toddler.

Elasmobranchs aren't alone on this branch of the fish evolutionary tree. Roaming the deep oceans are fish with rabbit-like heads, small mouths and nibbling teeth, and their bodies taper to a point, with a tail trailing behind like a ribbon. They're sometimes known as rat tails, or rabbit fish, but are more commonly called chimaeras.* Around

* They are the sole surviving order of the Holocephali, a sub-class with dozens of fossil species. Together with the elasmobranchs, they form a class called the Chondrichthyes.

420 million years ago, during the Silurian period, this sister group broke away from the elasmobranchs. Many chimaeras have bizarre head adornments, including noses that look like someone grabbed them and gave them a good tweak. Male chimaeras generally have a retractable organ on their head, which they use during sex; it has an opposable tip which slots into a notch on the female's head, stopping her from swimming off just when things start to get interesting.

Along America's Pacific Northwest coast, divers venturing out at night might glimpse a chimaera commonly known as the Angel Fish. They have big viridescent eyes, shimmering bronze skin covered in white spots, and they swim through open water in slow, corkscrew turns with flicks of their triangular pectoral fins. A few years ago in Puget Sound off Washington State, researchers found a pure white, albino chimaera, a very rare un-pigmented find for any fish – a real angel.

As we approach the base of our fish evolutionary tree there are just two groups left, but it's not a simple case of arranging them one after the other. This remains, in fact, one of the most controversial parts of the tree, with implications for the whole of vertebrate life.

These last fish groups look similar to eels (although all the way down here on the tree they're only distant relatives to those slender teleosts). They have a round mouth at one end and a flattened, paddle-like tail at the other, and both have something of a nasty reputation. Lampreys start out life harmlessly enough. All 38 species are born in rivers as larvae that spend several years buried in mud, filtering the water for morsels of food floating by. Then they grow into metre-long (3ft) parasites. Migrating out to sea, most lampreys hitch onto the skin of their host, often teleosts of some kind, and stick themselves firmly in place. They then set about scraping a hole in the host's skin with their sharp tongue before sucking blood or biting mouthfuls of flesh. The lamprey eventually unhitches itself and goes off to find its next victim; the host is depleted and may even die from its wounds.

Hagfish, the other group at the base of the fish evolutionary tree, have loose, scaleless pink skin that makes them look like they've crawled into a stocking. Compared to lampreys they're hardly any more genteel, and have the unappealing habit of eating a carcass from the inside out. You're likely to find one of the 70 or so species buried deep inside the decomposing body of a fish or a dead whale that's fallen to the seabed. They make an entry where they can, either through an existing orifice or by ripping a hole, then settle in for a feast, leaving behind nothing but skin and bones.

Two odd habits set hagfish apart from other animals. First is their legendary ability to make extravagant amounts of gooey slime. Put a hagfish in a bucket and soon you will have a bucket full of transparent ooze, squeezed from a series of pores along its body. In 2017, a truck transporting 3.4 tonnes (7,500lb) of live hagfish overturned in Oregon, smothering the highway in stringy, white slime. It took emergency services hours to clear up, with high-pressure hoses and a bulldozer, while thousands of hagfish were still slithering around. The intended destination for them was Korea, where hagfish are eaten and their slime used as a cooking ingredient, as an alternative to egg white. Researchers are also busy studying hagfish slime with the idea of making new materials and fabrics from the stretchy, thread-like proteins.

The slime is thought to deter hagfish-hunting predators by clogging their gills. In fact, if they're not careful, hagfish can quite easily suffocate themselves. To avoid this they tie themselves in knots – their second clever trick – and slide the knot along their bodies to rid themselves of their own goop. They'll do the same thing if you grab hold of one, using a knot to push against your hand and release your grip.

Traditionally, lampreys and hagfish have been distinguished from the rest of the fish based on the features they *don't* have. None of them have jaws. They're the only surviving jawless fish (many other jawless fish are known, as we'll see later, but they're long extinct). They also don't have complete backbones. But they do have a skull made of cartilage, and a stiffened nerve cord running along their back, the notochord. Consequently, they're considered to be the two most ancient groups of fish. Which of them evolved first, lampreys or hagfish, may not seem especially important, but this question lies at the heart of a great evolutionary debate.

It had been widely thought that hagfish kick-started the entire vertebrate lineage, some time around 500 million years ago. Recent genetic studies, though, support an alternative view that hagfish aren't the most ancient fish but are sisters to the lampreys. This view places the tight-knit duo together on their own branch of the vertebrate evolutionary tree.

This leaves us with a distinct gap between vertebrates and invertebrates, the animals with no backbones. The vertebrates' closest spineless kin are tunicates, also known as sea squirts. As adults, they sit about in the sea, stuck to reefs and rocks, quietly filtering water. It's in their younger stage, as larvae, that sea squirts reveal their vertebrate affinities, in the form of tadpole-like wrigglers, swimming around with a stiff notochord along their backs. This makes sea squirts firm members of the chordates – the major division of the animal kingdom within which the vertebrates sit – and they're what we would find if we carried on our journey along the fish evolutionary tree to the next branch down.

The search is on for the animals that came between the sea squirts and vertebrates. What did those first vertebrates look like, the common ancestors of hagfish, lampreys and all the other fish? These are the real missing links in the vertebrate evolutionary tree, right here at the base. We can look up into the branches swaying above us, laden with so many thousands of species: fish with skeletons made of bone or bendy cartilage; fish that live in mountain streams and at the bottom of the deepest seas; fish with lungs and fish with legs. This tree is also occupied by every other vertebrate, from whales and dolphins that went back to the sea, to the people who do their best to be amphibious, scuba tanks fixed to their backs. For now, though, it's still not clear how exactly this great lineage began.

How the flounder lost its smile

Isle of Man, traditional

A long time ago, in the sea near the enchanted Isle of Manannán, all the fish came together to decide who should be king. Each one hoped it might be them, so they all smartened themselves up and came looking their very best.

There was the Red Gurnard, Captain Jiarg, dressed in his fine crimson coat. Grey Horse the shark was there, big and fierce as always, with his skin polished to a shine. The Haddock known as Athag was there too, still trying to rub away the black spots that the devil had burnt on his skin.

Brae Gorm the mackerel swaggered about, certain that he would be king. He dressed himself in fine stripes of all the colours of the sea and sky, looking like a coat of diamonds. But the other fish didn't like his bragging and his gaudy costume, and they turned their backs on him.

Instead of the mackerel, it was Skeddan the herring who became king of the sea. While all the fish celebrated, another arrived who had also hoped to be king, but he was much too late. It was the flounder, known as Fluke. 'You've missed the tide,' all the fish shouted. 'Skeddan is now king!' The flounder had taken too long getting ready, decorating himself in red spots. 'Then what am I to be?' he cried. Scarrag the skate replied, 'Take that!', and with his tail slapped the flounder, knocking his mouth into a crooked frown on one side of his face – and so it has been ever since.

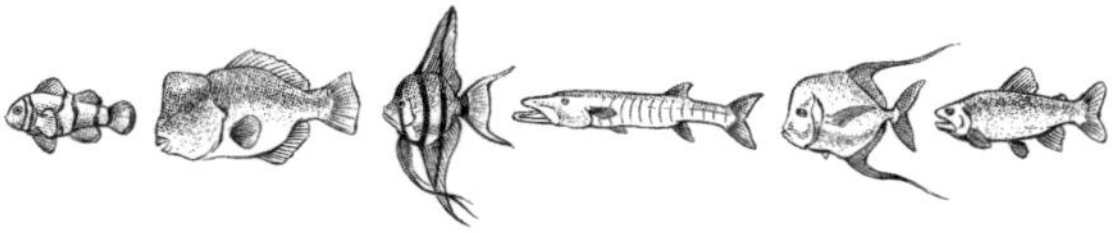

CHAPTER THREE:
OUTRAGEOUS ACTS OF COLOUR

CHAPTER THREE

Outrageous acts of colour

Stripes of inky blue and ripe banana-yellow run across the fish in front of me; its eyes are hidden beneath a black mask, like a gem thief. Emperor Angelfish somehow manage to be flamboyant and demure at the same time, which I expect is one reason why I like them so much. It's big for an angelfish, roughly as long as my fingertip to my elbow, and it slides under a dark ledge before spinning around to watch me from safety.

It's a species I always hope to see, and whenever I do I feel a contented sense of them always being out there, somewhere in the world, even when I'm not looking. I've seen Emperor Angelfish in the Red Sea, in the Maldives, the Philippines, Australia and Fiji; the same faces in different places.

This is Rarotonga, a mountainous, forest-clad island in the South Pacific encircled by a coral reef and a clear, turquoise lagoon. At high tide, I wade out from the beach – no boat required – and plan to stay as long as the tide will let me, threading my way around coral boulders, watching fish.

The lagoon is brimming with colourful life, like a well-stocked aquarium but without glass walls. I paddle around, taking in as much as I can. At first sight coral reefs can be an overwhelming scramble of colours and shapes, so dazzling and busy it's hard to make any sense of it all. But there are tricks for spotting fish and for working out what's what.

Start by looking at the shapes of fish. Common families have distinct profiles that can help you distinguish a bullet-shaped wrasse from an oval damselfish, a stout, wide-tailed grouper from a slender, fork-tailed fusilier. You can learn to pick out key characters like a soldierfish's big eyes, a goatfish's whiskery barbels hanging from its chin and the forehead of a unicornfish. With a search pattern in mind, you'll begin to notice groups of fish with similar shapes. They also behave and move in certain ways. Damselfish are the little ones, hanging about in flickering shoals over colonies of coral, darting between the coral's branches and fingers when you come near. Blennies usually hide away in holes in coral boulders, and occasionally pop their heads out. Gobies look similar to blennies, but they can be bigger and often live in burrows on the seabed, together with a shrimp partner; the shrimp industriously shovels gravel out of the burrow while the goby watches out for trouble. Wrasse and parrotfish row themselves through the water with beats of their pectoral fins; triggerfish swim by undulating their large dorsal and anal fins, on their back and belly; cardinalfish hang motionless close to the reef.

Colours and patterns can then narrow things down and help you to identify particular species. Some of the easiest to pick out are butterflyfish and angelfish, which have bold colours, spots and stripes. Panda Butterflyfish are unmistakable, with their dark eye-patches, as are Raccoon Butterflyfish, with white and black bandit eye-masks like their furry namesakes. Keyhole Angelfish are deep blue all over except for an oval white patch, which you might imagine you could put your eye to and peer through.

As you get to know fish and start spotting your first few species, you'll learn to make mental notes of things you don't recognise to look up later. I do this in Rarotonga's lagoon. There's a white and yellow butterflyfish with a dark blot that's smudged as if it got caught in the rain, and an angelfish, yellow all over, with neon blue edges to its fins and a ring around each eye like a pair of spectacles. Later, I flip through my identification guide and learn these two are Teardrop Butterflyfish and Lemonpeel Angelfish.

Sometimes there are rare locals to look out for. In Rarotonga, just in case, I keep an eye out for an angelfish with a scarlet body and five minty white stripes running across it. This is the Peppermint Angelfish, discovered here in 1992 on a deep reef and never seen anywhere else in the world. A living specimen was collected during a research expedition run by the Smithsonian Institution in Washington DC and donated to the Waikiki Aquarium in Hawaii, so researchers could study it and members of the public see it. The aquarium keepers turned down numerous offers from trophy-hunting private fish collectors of up to $30,000 for this singular little fish. But I'm kidding myself, really, that I'll see one. Peppermint Angels have only ever been seen much deeper, far beyond the reach of my lungful of air.

The tide begins to fall and I dawdle reluctantly back towards the beach as the water drains from the lagoon. Bright fish still dart about, right in front of my mask. Many are adolescents that will change colour as they grow up. Among them I spy a rare treasure that makes me smile, sending a dribble of water into my mask and up my nose. It's the shape and size of a large blueberry, only canary yellow with black polka dots and a pouting snout, and it's bobbing restlessly left and right. It could be a tiny helium balloon being dragged behind a restless child, but it is in fact a young Yellow Boxfish. When it gets older, it will lose its bulbous profile, become more box-shaped and swap its bright yellow outfit first for dirty mustard, then for blue.

Picasso Triggerfish also patrol the lagoon, the size and shape of a flattened rugby ball. Airbrushed patterns adorn their flanks with subtle shades of yellows and tawny browns. It looks as if an artist then dragged his fingers through the wet paint, leaving four white stripes. Between the fish's golden eyes run bands of ultramarine; dripping down its face are iridescent blue tear streaks, and on the upper lips a pencil moustache to match. Picasso Triggerfish hatch in this audacious birthday suit, which they'll keep throughout their lives. Finally I stand up and walk to the beach and I see minute painted Picassos, their patterns shrunk down to a tiny size but still unmistakable, as they zip around in a finger's depth of water.

Colourful fish don't only live on coral reefs. When adult Coho Salmon leave cold North Pacific waters to spawn in forest-lined rivers, they transform from silver to crimson. On Britain's south coast I've seen turquoise spots gleaming in a tide pool, like a pair of eyes gazing up at me, from the back of little fish called Cornish Suckers. In North America, along the coast between San Francisco and Baja California, bad-tempered fish called Sarcastic Fringeheads live inside big, empty seashells. Males charge at each other and engage in irate head-to-head contests, opening their enormous jaws like umbrellas that are yellow and red on the inside. Further east, streams and cool, clear creeks across the Mississippi Basin are home to almost 200 species of darter. Some of these finger-sized fish live only in a single stream and most are distinguished by their bright patterns. Candy Darters are iridescent jade striped with orange; Banded Darters come dressed in bright emerald stripes. A 2012 study revealed that the Speckled Darter is in fact at least five separate species with distinct colours. The newly recognised darters were named after US presidents did their bit for who environmental protection; among them, the Spangled Darter *Etheostoma obama* is vivid orange, with blue spots and stripes.*

* As part of Barack Obama's legacy, he expanded the Papahānaumokuākea Marine National Monument in the central Pacific, protecting 1.5 million square kilometres (580,000 square miles) of ocean waters, including many coral reefs and the home range of the critically endangered Hawaiian Monk Seal.

Why fish are so colourful is a question many ichthyologists have pondered. And through their studies they've seen that fish expertly use colour as a tool – to hide and shock, warn and woo. Fish adapt to the way light and colours behave underwater, as sunlight splits apart and wavelengths shift and change. Studies of colourful fish are also revealing broader details of how the living world works. Fish colours are showing the startling pace of evolution and they offer up clues as to how the polychromatic world around us came to be.

In December 1857, naturalist and collector Alfred Russel Wallace arrived on the Indonesian island of Ambon, where he hired a boat to cross the bay and reach the island's interior. He was part-way through an eight-year, 22,000km (14,000-mile) journey around Southeast Asia, during which he gathered tens of thousands of animal specimens, made detailed observations of people and wildlife and, independently of Charles Darwin, formulated a theory of evolution. The water in Ambon's bay was so clear that Wallace could see all the way down to the seabed from the surface. Without even sticking his head beneath the waterline, he saw for the first time a coral reef. It was, he later wrote in his book *The Malay Archipelago*, 'one of the most astonishing and beautiful sights I have ever beheld.'

The hills, valleys and chasms of what Wallace called 'these animal forests' made of coral and sponges were inhabited by shoals of fish, 'blue and red and yellow … spotted and banded and striped in a most striking manner … It was a sight to gaze at for hours.'

Twenty years later Wallace wrote an essay exploring ideas of why many living things are brightly coloured. For species to daub themselves in such showy pigments would seem at first to be a dreadful idea. Eye-catching displays are surely an open invitation to passing predators to come and dine. Equally, the hunters should conceal themselves as they sneak up or swoop down on their prey. Nevertheless, dramatic colours have evolved time and again throughout the natural world, and especially among fish. In Wallace's view, this mostly comes down to animals trying to keep themselves safe.

'Even gay colours are very often protective,' Wallace wrote, 'because the earth and the sky, the leaves and the flowers, themselves glow with pure and vivid hues.' Living in such a colourful world, it follows that animals might use bright colours to hide themselves. There are green caterpillars with pink spots that resemble the flowering heather they like to nibble. Wallace pointed out that the green feathers of many tropical birds, from parrots to white-eyes, bulbuls, barbets and bee-eaters, match the evergreen foliage they inhabit. And in polar regions white is the colour of choice, for Polar Bears and Arctic Foxes, all the better for hiding in snow and ice.

Reef fish use their colours as camouflage, Wallace proposed, to hide among bright seaweeds, anemones and corals. The examples he gave included the seahorses and their Australian relatives, with 'curious leafy appendages', the Leafy and Weedy Sea Dragons. Since Wallace's time, divers and scientists have discovered many other well-camouflaged seahorses, with pink and orange pimples that match the knobbly sea fans where they perch, or yellow and purple tufts like their soft coral homes.

The Warty Frogfish is another reef species with immaculate camouflage. This dumpy fish sits, mostly unmoving, and alters its colour and texture to match its surroundings, often yellow and orange sponges. It's the kind of fish that's so well hidden you need two fingers pointing from different directions to show another diver where it is. In 2016, a pure white frogfish was spotted in the Maldives. This was shortly after the reefs were hit by a major coral-bleaching event, triggered by high water temperatures. Heat-stressed corals expel the tiny, pigmented algae that live in their transparent tissues, so they lose their colour and become ghostly white. The frogfish had quickly responded to this change in its world; it had even sprouted small green tufts, apparently to match the seaweeds that were beginning to grow over the dead white coral.

For slow-moving fish like seahorses and frogfish this form of colourful camouflage can work astonishingly well. But what about more active fish? The scene behind them is constantly changing and calls for more sophisticated camouflage. A 2015 study revealed how Slender Filefish from Caribbean reefs perform lightning-fast costume changes. Second by second, a filefish choses from at least 16 different outfits to coordinate with whatever it happens to be swimming past: a patch of green seaweed, a pale, lacy sea fan or a golden frond of soft coral. These little fish are always dressed for the occasion.

Some fish even dress up like their prey. In the waters around Lizard Island on the Great Barrier Reef live small, predatory fish called Dusky Dottybacks, a single species that can be either yellow or brown. They hunt for young damselfish, of either a yellow or a brown species. A research team, lead by Fabio Cortesi from the University of Basel in Switzerland, conducted a neat experiment to test how these predators use their colours. Divers stocked an experimental area of wild reef with either yellow or brown damsels, then in various combinations they added the predatory dottybacks, again either the yellow or brown colour morphs.* After two weeks, the divers checked on the fish and saw that the predators had changed colour to match their prey. Further studies in the lab showed that the

* Morphs are variations of the same species within a population that somehow appear different, in size or colour, for example. Often it's males and females that differ from each other, with this phenomenon known as sexual dimorphism.

predators had more success when they were the same colour as the fish they were hunting. By matching the damsels' colour, the predatory dottybacks can sneak in and get closer to the young fish, who are less wary and presumably mistake the dottybacks for adults of their own species. It seems the dottybacks are performing a fishy version of a wolf in sheep's clothing.

Colours and patterns also help fish hide in plain sight, by breaking up their outline. The blue and yellow stripes of Emperor Angelfish, rather like a zebra's stripes, could make it difficult for predators to pick out the shape and target of a fish. In shoals of stripy fish it's tricky to see where one fish begins and another ends. Fish also use their patterns to hide particular parts of themselves. The Emperor's dark eye-masks could reduce the chances of predators pecking at their eyes.

Butterflyfish often combine eye-stripes with large, dark spots elsewhere on their bodies, often ringed in iridescent blue. It's thought these could be false eyes, diverting attacks away from a fish's delicate head-end. Eye-spots may also confuse predators when their prey swims off in apparently the wrong direction.

Despite decades of research, however, these theories are proving tough to test and there's still no consistent view for why so many different animals have evolved eye-spots, from fish to birds, butterflies and moths. In another study of damselfish living around Lizard Island, researchers didn't find any juveniles with bite marks near the spots on their tails, suggesting the markings aren't confusing predators into attacking the wrong end. Monica Gagliano from James Cook University in Queensland, Australia, proposed that the spots could be 'decorative leftovers' from some past survival advantage that no longer applies. Or maybe predators more recently got smart to eye-spots and are no longer fooled.

In his essay on protective colours, Alfred Wallace noted that nocturnal animals tend to be dusky and dark to blend in with the shadows of the night. By contrast, the common colour for underwater night-dwellers isn't black but red. There are red squirrelfish, soldierfish and bigeyes, all of them tropical, nocturnal species. Red is also a common costume for animals living in the deep. A newly discovered species of the seahorses' relative, the sea dragon, was recently filmed for the first time deep underwater off the coast of Western Australia. It has skin the colour of rubies. On deep underwater mountains live Orange Roughies; alive they're brick red, fading to orange when they die. All these fish evolved their ruddy complexions because of the way sunlight behaves in water.

When light from the sun reaches the Earth, it's composed of a blend of colours, each with a certain wavelength. Humans can generally perceive colours ranging from short-wave blues and purples to longer-wave oranges and reds. All wavelengths of light pass with equal ease, more or less, through air, and when we see them all together our brains interpret them as white. Passing through water, however, sunlight begins to separate out, as longer wavelengths, with less energy, are quickly absorbed by water molecules. It means that in clear water below 20 metres (65ft), very little red light remains. Go deeper and other colours blink out; oranges are lost, then yellows and greens. Blue light, with its energetic, short wavelength, penetrates the deepest. This is why most of the oceans are blue. It follows that lots of creatures inhabiting the open sea evolved to be blue, to match their surroundings.

Why, then, are deep and nocturnal animals red when that isn't the colour of the sea itself? The reason lies in the way coloured pigments work. Most objects appear a certain colour because they contain pigment molecules that absorb specific wavelengths of light and reflect others – and it's the reflected colours that we see. Hence a leaf in autumn looks red because it contains anthocyanin pigments that absorb green and blue light, and reflect red. However, if you were to take that red leaf on a deep dive it would quickly fade to inconspicuous grey then black because there'd be no more red light available to reflect. The red pigment would be unable to show its true colour. Similarly, as the sun sets, the first wavelength to ebb away underwater is red, making this a useful colour for both nocturnal and deep-water camouflage.

In shallower waters, during the day, red can be a good colour not for hiding but for being as obvious as possible. Alfred Wallace wrote about the colours and patterns that serve as a warning to other animals. Bright colours can call attention to hidden poisons or dangerous spikes and stings that attackers do well to learn to recognise and avoid, like the black and yellow stripes of bees and wasps. Wallace didn't refer to such cautionary colours in fish, but there are plenty. Lionfish have red and white stripes, warning of their long, venom-tipped spines. Surgeonfish are named after the venomous blades at the base of their tails, which are often bright warning colours.

Some colourful fish paint themselves in warning colours when in fact they're quite harmless. Innocuous species gain protection by wearing false warning colours and pretending to be poisonous. One such mimic is the Whiteblotch Sole which, in its young stage, masquerades as a toxic flatworm. The fish and the flatworm look remarkably alike. Both lie flat on the seabed and ripple slowly along, their black bodies fringed in orange, with large white spots. Mistaking it for a bad-tasting flatworm, predators should leave the sole well alone.

Wallace concludes his musings on animal coloration discussing what he calls 'one of the most curious chapters in natural history.' He considers how colours are passed on and perfected from generation to generation. Pigments and patterns vary within a population of animals and some will give individuals a 'better chance of life', as he puts it, by making them less conspicuous or by warning off enemies.

Those most useful colours will pass from parents to offspring; the same selective process continues in each generation until camouflage or warning colours become so good that enemies give up and focus attention on other species. But it doesn't stop there. Over time, predators may change too, and get better at seeing through the deception, so generations of prey will continue to adopt the best, most protective colours and patterns. Wallace didn't label it as such, but this is evolution in action. It leaves us, he writes, with a satisfactory clue as to why we see such 'varied coloration and singular markings throughout the animal kingdom, which at first sight seem to have no purpose but variety and beauty.'

Wallace's ideas explain the bright colours of many animals, but not all use colours to be secretive or scary. There are fish that aren't hiding or poisonous, or even pretending to be poisonous. Since he wrote his essay on protective colours other theories have emerged, including the possibility that fish are writing colourful messages on their bodies intended for other fish to read.

Fifty years ago, Austrian zoologist Konrad Lorenz came up with the idea that many coral-reef fish decorate themselves with what he called *plakatfarben*, or poster colours. He proposed that they use their bodies as a living billboard, adorned in brash adverts declaring their identity and gender.

Lorenz, a pioneer in studies of animal behaviour, was especially interested in aggression. He kept coral-reef fish in aquaria and watched how they got along – often not at all well. Fights frequently broke out between members of the same or similar species. They would nip, bite and eventually kill each other. To see what happened in a more natural setting, Lorenz went to the Florida Keys and Kaneohe Bay in Hawaii. Snorkelling with wild fish, he saw similar conflicts playing out, although usually not fatally; rather than hang around and get a beating, the losing fish would quickly swim away. Their bright colours and patterns, he thought, were telling fish who was who. This is similar to the team colours sports fans wear, except that fish aren't looking for a fight with rivals but with members of their own team. Other fish of the same species can be their greatest competitors. Lorenz became convinced that fish use their bright displays to stake out territories and fend off intruders, nailing their colours to the reef.

Lorenz's poster-colour theory has been supported but also refuted by various subsequent studies. Some fish seem to fit the model that brighter colours belong to more argumentative, aggressive species. Red-toothed Triggerfish, for example, are coloured a relatively inconspicuous deep blue all over, and they are quite a gentle species; meanwhile, Picasso Triggerfish, like the ones I saw in Rarotonga, are far more ostentatious in their colours, and more quarrelsome. There are, however, other species in which the opposite is the case: angry fish with drab colours.

A further strand of evidence in favour of poster colours comes from the fact that in many species, adults and juveniles look completely different. In their early years Emperor Angelfish are dark navy, with white and electric blue concentric rings across their bodies. Only when they're around two years old will their yellow and blue stripes gradually seep through their skin, replacing the rings. It's thought this costume change allows younger fish to appease the adults.

In a 1980 study of Emperor Angelfish in the Red Sea, German biologist Hans Fricke went diving and took down with him two small painted wooden fish, one with adult stripes, the other with juvenile rings. He fixed them to the reef in the territories of various emperors and watched their response. You might think the real fish wouldn't be fooled by a non-moving, wooden replica but they did seem to react differently to the two patterns. Adult emperors were more likely to attack the model adult fish and generally left the young one alone. This simple study suggests that the young fish's colours may placate truculent adults and win them safe passage across the reef, until such time as they're ready to make a bid for their own territory and reveal their mature colours. Several other angelfish species wear a school uniform similar to the emperors', and the species can be difficult to tell apart when they're young. Again, this lends support to the idea that young fish are trying to avoid getting into fights.

This doesn't seem to apply to all fish, though. In another study, this time of large damselfish called Garibaldi that live in Californian kelp forests, the juvenile colours seem to offer little protection. Thomas Neal from the University of California (Santa Barbara) gathered various young Garibaldi, some still peppered with the metallic blue spots of their youth and some that had recently lost them and were orange all over. He presented these similar-sized but differently coloured juveniles to adults and watched as they launched aggressive attacks on the fish with spots.

Rather than chasing off intruders, some poster colours invite them to stay. Various species of wrasse and gobies take on the role of dentists and hygienists on coral reefs. They set up cleaning stations and spend their days pecking away dead skin and scales from client fish, cleaning between their teeth and pulling blood-sucking parasites from their skin. Many of these cleaner fish have blue and yellow stripes, a common colour combination on the reef. These two colours can be seen from the furthest away in clear, blue waters. And being widely separated on the colour spectrum, blue and yellow contrast strongly underwater. Wearing these colours, cleaner fish advertise their services far and wide.

Theories that fish use colours to hide themselves away or to shout as loudly as possible rest on the assumption that they themselves can see these colours. And indeed they can. Fish have eyes with a similar basic structure to our own. Like us, they have a pair of liquid-filled orbs with a narrow pupil to let in light and a lens to focus an image on the retina, a layer of light-sensitive cells at the back. One difference between fish and human eyes, though, is the lens shape. As light passes from air into our eyes, it already begins to bend inwards and focus towards the retina, because the refractive index differs between air and liquid. Muscles then adjust the shape of our elliptical lens to fine-tune the image (except for those of us who are long or short sighted and need glasses to help). Open your eyes underwater and everything becomes blurred because you've instantly lost that focusing power between air and eyeball, and light passes straight in. If fish had lenses shaped like ours, they'd have to wear incredibly thick spectacles to see clearly. Instead, they have spherical lenses inside their eyes, which bend light much more strongly. Next time you cook a whole fish for dinner, dig into its eye and you'll find the lens, neat and round like a ball bearing, and opaque once you've cooked it, because the proteins inside have been denatured like those in a boiled egg. When focusing on things near or far away, fish move their whole lens around inside their eyeballs, like moving a magnifying glass closer or further from your eye.

Colour vision comes courtesy of specialised light-sensitive cells in the retina known as cones. Each cone responds to a specific range of wavelengths; they absorb photons of light and fire nerve signals to the brain. By comparing signals from different cones, the brain interprets colours. Humans generally have three cone types, and our brain translates their firing as a continuous rainbow of colours between indigo and red.

To determine the colours fish can see, biologists dissect out their retinas and use machines called spectrophotometers to beam light on them and measure which wavelengths they absorb. Since the 1980s there have been micro-spectrophotometers, which shine thin needles of light onto individual cone cells. Studies like these have shown that, between them, fish have a whole range of cones. Some species have two, others have four, and some are sensitive to wavelengths that humans can't see.

For example, various freshwater fish have evolved red-shifted vision. They can see far-red and infrared light, which we can't. The reason for this is that as sunlight filters through freshwaters, flecks of mud and algae absorb certain wavelengths and push ambient light towards the red end of the spectrum, so there's more red light to see by. Not only that, but some migratory species change the colours they see best as they move inland from the sea. Salmon and lampreys see blue light better while they're swimming out in the blue oceans. Then, when they move inland, they adjust their visual pigments to see far-red and near infrared light. Lemon Sharks change their vision in the other direction. As youngsters they inhabit murky waters, among the roots of mangrove forests, and later in life move offshore, adjusting their vision from red to blue as they go.

There are fish that can see ultraviolet light. It had long been assumed that fish wouldn't be able to see UV since it gets scattered and noisy underwater, making it of little use. But it turns out that for short-lived species, UV is the perfect wavelength to send out secret, close-range messages that other fish, mostly predators, are blind to. Studies of damselfish reveal they recognise each other and distinguish species from intricate patterns on their faces that reflect UV light. Two species of little yellow damselfish, Ambon and Lemon Damsels,* are almost identical to the human

---

* Ambon Damselfish were named in 1868 by the Dutch ichthyologist Pieter Bleeker, a decade after Alfred Wallace was in Ambon; the species ranges across the Western Pacific from Japan to Australia.

eye except their UV facial patterns differ. These patterns seem to act as hidden poster colours that damsels can see but predatory fish generally can't. Predators tend to be longer-lived and have UV filters in their eyeballs – their own inbuilt sunglasses – probably to protect them from years of sun exposure. The smaller, short-lived species seem to take advantage of this, and use UV decorations to communicate with each other without being spotted by their enemies.

Fish produce their dazzling, vibrant colours with specialised cells in their skin called chromatophores. Pigment granules give these star-shaped cells particular colours. Common types include black melanophores, red erythrophores and yellow xanthophores. Extremely rare are cyanophores, the cells with blue pigments. So far they've only been detected in two animals, both of them fish: the Picturesque Dragonet and its close relative, the Mandarinfish.[*] I've spotted a Mandarinfish, the size of my little finger, peeping from a branching coral in a shallow lagoon in Palau, in the western Pacific. It briefly showed me its splendid green and orange body patterns, and wide fins trimmed with intense ultramarine, like powdered lapis lazuli.

Apart from these two blue fish, all the other blues in the living world are not pigments but structural colours. Instead of simply reflecting particular wavelengths, as pigments do, structural colours are produced when light bounces around inside a material and gets reflected, diffracted and scattered in different ways. There are structural colours throughout nature, from blue skies, rainbows and blue eyes to a butterfly's wings and a Vervet Monkey's brilliant blue

[*] It's likely that other blue pigments will eventually be found in other living things. Blue-skinned poison-arrow frogs are candidates, but no one has yet been brave enough to check.

scrotum. The common silvers and blues in many fish are structural colours made by another type of skin cell called iridophores. These contain guanine crystals and act like tiny mirrors, reflecting and interfering with the light that falls on them. They gleam in a similar way to the nacre (or mother-of-pearl) made by molluscs, and indeed fake pearls have often been made by covering glass beads with ground-up fish scales.

Layers of chromatophores and iridophores combine to produce all the fish's colours and decorations. These can change, gradually or within split seconds as muscles alter the orientation of crystals and squeeze the chromatophores tight or stretch them out, hiding or revealing the pigments inside.

In open waters many silvery fish use layers of iridophores to disappear from sight, even though there's nothing to hide behind. Anchovies, herring, mackerel and tuna take advantage of the unique light conditions underwater to camouflage themselves. To grasp how this works, first imagine a sheet of clear glass hanging vertically in open water. If you look up at the glass from slightly below, it's invisible because light passes straight through and all you see is the water behind it. By contrast, out of water, you can usually see a sheet of glass because some light reflects off it and back at you, which doesn't necessarily match the background; you might see yourself in the glass. Now, back underwater, replace the glass pane with a mirror. Unless a shoal of fish swims right past and you see their reflections, the mirror disappears just like the glass did. All you see is blue light that exactly matches the background. This happens because as daylight seeps down through water, it fades in a uniform way and is the same intensity around a horizontal plane; if you spin slowly around underwater, the brightness of water you're looking at won't noticeably change. The key thing is that the submerged mirror reflects

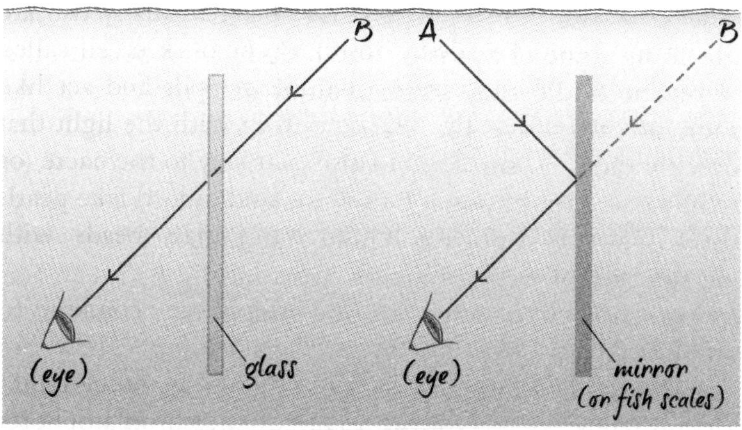

How a mirror, or a silver-sided fish, behaves like a pane of glass and disappears from view underwater.

the light in front of it, and this is of the same intensity as light behind it, on the other side of the mirror (in the diagram, light beam A matches light beam B). It's impossible to know if you're looking at a reflection in a mirror or peering through a pane of glass at the patch of water behind it – the two appear just the same. In this way, fish can also vanish underwater by covering their bodies in mirrors, so they blend in with the water around them. For this to work they have to keep their silvery bodies vertical. This is probably one reason why many species are incredibly thin and compressed from side to side. Fatter fish get around this by arranging stacks of crystals vertically within their skin, even on the rounded parts of their bodies, creating the same effect as a flat mirror. But still fish move and deviate from vertical, breaking their disappearing act, which is why silvery schooling fish glint and flash as they swim around in spirals.

The best way to appreciate the bright pigments and gleaming shine of fish is to watch them from beneath the waterline while they swim around in front of you. As soon as they leave the water, fish's colours and lustre quickly fade; no surprise, really, given that they evolved to be viewed underwater. Before underwater photography, the only way to capture and hold onto their brilliance was via the skilled work of artists, who ideally had seen the living fish themselves, sometimes in remarkable circumstances.

In March 1790, the British Naval ship HMS *Sirius* struck a coral reef next to Norfolk Island, almost 1,450km (900 miles) off the east coast of Australia. It was the flagship of the First Fleet, the ships that had come to set up a penal colony in Australia a few years earlier. When the ship foundered, the first thing Captain John Hunter did was to make sure nobody drowned. All 200 people, mostly British convicts, were taken safely ashore. Over the next few days as many provisions as possible were rescued from the hold, before the ship broke apart. Among the belongings that came to shore was a paintbox belonging to midshipman George Raper.

Since he left Britain aboard the *Sirius* three years previously, Raper had been painting charts and ports along the route to the southern hemisphere. Shipwrecked and facing dwindling supplies and starvation, Raper nevertheless turned his attention to painting the local wildlife.

When rescue eventually came 11 months later, Raper brought with him a collection of intricate, colour paintings including many fish from the waters around Norfolk Island. There's a Sweetlip Emperor, with scarlet fins and yellow lips, and a Sandager's Wrasse, with purple and green blush on its cheeks; both are easily recognisable as species that still live around the island today.

Other fish artists weren't quite so true to life.

Earlier in the 18th century, on the island of Ambon in Indonesia where Alfred Wallace would later marvel at the coral reef, a Dutchman spent several years drawing and painting fish. Samuel Fallours served as a soldier and then a clergyman's assistant in the Dutch East India Company. Local fishermen brought him fish that became the subject of paintings which Fallours sold to Company officials and European collectors. His elaborate artwork was reproduced in several books, including *Poissons, écrevisses et crabes* (Fish, crayfish and crabs), published in the Netherlands in 1719. It was the first colour book of fish and only a hundred copies were ever printed, making it one of the world's rarest natural history books.* The pages are filled with a stunning menagerie of strange creatures. Fallours's drawings are highly stylised, more artwork than meticulous illustration, but still it's possible to work out many of the species he drew. There are boxfish and triggerfish, lionfish and butterflyfish. Certainly he was more creative with his paintbox than George Raper, perhaps deliberately, to sell more pictures to his European clients who avidly collected exotic novelties. Stripes and spots are bold and exaggerated; bodies are filled in with invented geometrical patterns and ornamental twirls. If there had been an aquarium tank in Wonderland, these are the animals Alice would have watched through the glass.

On real fish species, extraordinary acts of colour often come about because females prefer the brightest mates. Males strike flamboyant displays. *Choose me, choose me*, they quietly shout. *I am quite the best father you'll find.* You can see the signature of sex written across many animals' bodies, usually in the differences between males and females. In

* In 2007, a copy sold at auction in London for £43,200.

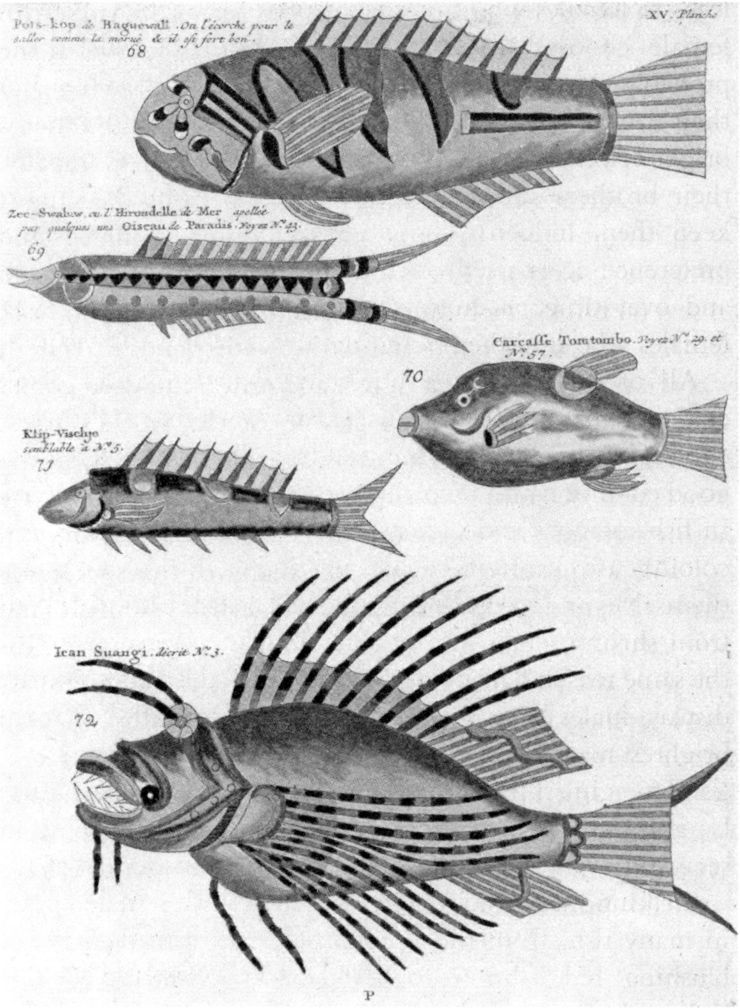

Various fish originally by Samuel Fallours, possibly wrasse, pufferfish and lionfish, from Poissons, écrevisses et crabes, *1719.*

general, females are sombre and inconspicuous, while males are far more splendid and eye-catching. The peacock's plumage and the Mandrill's blue and pink rump are prime examples. These and many other gaudy male characteristics become exaggerated over time by the mingling of two types of genes: one produces colourful attributes, the other leads to females who find those colours attractive. When a female chooses to mate with a colourful male, she'll probably produce either male offspring that are vivid like their dad or females that share her penchant for those same bright colours. The girls don't need to show off as much as their brothers and they'll carry the colourful genes, but keep them hidden. Those genes for colour and colour preference are passed on together down the generations and over time produce ever brighter males, along with females who find them fetching.

All of this raises the question of why females prefer colourful mates in the first place. Far from being a whimsical choice, bright colours can in fact be a clear sign of a male in good condition and a worthy partner with top-notch genes. In fish, oranges and reds are especially good signs. These colours are carotenoid pigments that fish can't make themselves and so come exclusively from their food, mostly from shrimp, crabs and other colourful invertebrates (it's the same reason flamingoes are pink). To put on a colourful display, males have to eat a lot of food. It follows that the brightest males are well-fed, healthy, strong swimmers with good foraging skills, all qualities that tend to be underpinned by good genes. By choosing colourful mates, females are securing a good genetic inheritance for their offspring.

Striking male colours are associated with female choice in many fish, from multi-coloured American darters and blushing red salmon to sticklebacks with red bellies, Zebrafish that court each other with brief bright flashes, and parrotfish that begin life as subtle females, then change sex and change colour to become flamboyant males. And there's one particular fish species that's shown how the watchful gaze of females can be an extremely potent force, and how sex can drive evolution in certain colourful directions. These fish have been given a few different common names over the years. They've been known as 'millions' because there are so many of them, and 'rainbowfish' because the males are covered in colourful spots, stripes and splashes. Most commonly, though, these little, 3cm- (1in-) long fish are known as Guppies.

They were named after an Englishman, Robert Guppy.* He came across the species 150 years ago, although he wasn't in fact the first person to find them. A few years earlier they'd been spotted by German explorer Wilhelm Peters. But it was Mr Guppy who brought these little fish to the attention of the English-speaking world, which is why we now have Guppies and not Peters.†

Since their discovery, Guppies have become some of the most widespread and cosmopolitan fish; they've been introduced to freshwaters worldwide as part of attempts to control mosquito larvae and stem the spread of malaria; they've spent time on the International Space Station; and they live in their multitudes in aquarium tanks as much-loved pets.

Originally, before people started moving them around and off the planet, Guppies were native across the Caribbean and into South America. Mr Guppy found them in Trinidad. Across the island's northern coast runs a mountain chain covered in lush cloud forests that are home to Howler Monkeys, Ocelots, Musk Hogs and critically endangered Golden Tree Frogs. Rivers and waterfalls tumble down from these cool, misty elevations, filling up clear pools and streams that are home to Guppies.

Throughout the last 60 years, a succession of biologists has scrambled through Trinidad's montane forests to study Guppies in their wild habitat. Early on, wife and husband team Edna and Caryl Haskins from the US were the first to notice that not all Trinidad's Guppies are equally colourful. In some ponds and streams they found males fluttering like rainbows, in others they looked more like females, with muted colours.

---

* Known by his middle name, Lechmere. Robert Lechmere Guppy.

† The Guppy's scientific name has shifted over the years, with various synonyms rejected in favour of *Poecilia reticulata*.

In the late 1970s, John Endler from Princeton University in New Jersey came to Trinidad and began photographing Guppies. His pictures revealed an intriguing sequence. Near the top of the mountains, when he dipped in pools, scooped out fish and took their photos, he saw the most colourful Guppies. As he worked his way downwards from pool to pool, the fish's colours gradually faded away.

Endler also noticed that other Trinidadian fish varied from place to place. In the higher ponds he found just a single predatory species, the Leaping Guabine. This gets its name from its habit of occasionally leaping from the water to snatch beetles and ants from overhanging vegetation. On rare occasions, Leaping Guabines will eat a few small Guppies. Lower down the mountains, and in certain deep valleys, Endler found a greater variety of predatory fish, including a voracious hunter called the Millet – the Guppies' most formidable enemy. Millets don't venture upstream because waterfalls and rapids form barriers, which means the headwater streams and pools remain a much safer place for Guppies, out of the range of these hungry foes.

As Endler contemplated the varying fish communities from the top to the bottom of Trinidad's mountains, a question arose in his mind. Are male Guppies treading a fine line between being seen and being eaten? On the one hand they need to be colourful enough to attract females; on the other, they can't be so colourful that predators easily spot them.

Using thousands of photographs and fish measurements, Endler revealed a beautiful correlation. Male Guppies are at their brightest in the safest ponds with the fewest predators; there they have lots of big spots, especially in blues and shiny conspicuous hues. Where more predators prowl, the males' colours are at their most neutral.

But, as every science student should know, correlation does not necessarily imply causation. It could be merely a coincidence that the Guppies' vulnerability and drab colours seem to be linked. To test his idea further, Endler performed some experiments.

In July 1976, he picked a dangerous (for Guppies) stream lower down in Trinidad's mountains, where Guppies lived alongside lots of Millets. He scooped out 200 Guppies, snapped pictures of them all, then released them in another stream not far away, one that was cut off by a waterfall. In their new home the Guppies cohabited only with Leaping Guabines, the moderate predators that pay them little attention. Essentially, Endler had switched off the predation pressure.

Two years later, Endler went back and scooped out Guppies from this translocated population and took their pictures. In the relatively short time they'd spent away from their main enemies, the Guppies' colours had intensified. It wasn't that the individual fish had changed colour, but that the brightest males had been the most successful in attracting females and had fathered lots of offspring, passing on their colours. Within just a few generations* the genes for bright male colours had spread and the average colour of new male Guppies became more vivid. Compared to the parent population they came from (which Endler also checked on), the liberated males had bigger, more colourful spots.

By manipulating the Guppies' predicaments, Endler uncovered vital clues as to how evolution works when multiple factors are at work – and unusually, he had done this in the wild and not in a laboratory. Females select for bright colours, and predators keep flamboyance in check; these two conflicting forces act on a population, shifting and changing over time. When predation is higher, males

* Female Guppies produce their first offspring at the age of 10–20 weeks; males mature after seven weeks or less. Unusually for teleosts, Guppies mate with each other: the male inserts sperm into the female with a modified anal fin, the gonopodium, and she gives birth to live young.

are most successful if they tone down their mating colours. And when conditions are safer, it pays for males to be as bright as possible, to impress females.

Not content with just his wild studies, Endler also took some Guppies back with him to Princeton. There he built a series of ponds to mimic Trinidad's streams. He made some of them relatively safe, stocking them with Leaping Guabines, while others were more dangerous, and stocked with Millets. Then to each pond he introduced a mix of Guppies.

After just 14 months he saw the male Guppies in safe ponds had the brightest colours. In dangerous ponds, the males' spots had shrunk and the blue and iridescent spots had disappeared. It was as if Endler had found the brightness button for the fish's colours, and he could turn it up and down at will.

In the streams of Trinidad, Guppies have nowhere to hide, and predation is an invincible force. The situation is different on coral reefs where the complex, rugged habitat allows fish to slip out of sight and conceal their colours. When predators are near, fish dash under a head of coral or into a small cave to hide. When the coast is clear they emerge and flaunt their colours, even positioning themselves in sunbeams to show off to mates and warn off intruders. And some reef fish spread out the pigment granules in their chromatophores, making their poster colours even brighter when they meet competitors. At night they dim their colours to reduce the chance they'll be spotted while they rest on the seabed.

John Endler helped to make Guppies some of the most famous fish in the world, at least among biologists. In the last few decades, they've become a popular species for researchers who are trying to answer big questions about evolution and ecology.

Guppies now live in colonies in research labs worldwide and biologists still venture out to study them in the forested mountains of Trinidad. Every year more details of their lives come to light, and with them come new insights into how flexible and adaptable living things can be. In 2007, researchers at the University of Toronto saw that male Guppies have uneven patches of colour on either side of their body – their patterns are not symmetrical. While dancing for a female, trying to convince her to mate with him, a male will show her only his better, more colourful side.

A 2013 study in Trinidad revealed that female Guppies prefer odd-looking males with rare colours. But like a new fashion that starts among a few, edgy individuals – like hipster beards, lumberjack shirts or thick-rimmed spectacles – it soon catches on until almost everyone looks the same. Eventually those colours that once were rare and exciting among the Guppies become too popular and fall out of favour. Females probably evolved a preference for rare colours because it helps them avoid inbreeding and mating with close relatives. As an upshot, Guppy populations cycle through colours, keeping a mix of many tints and maintaining their reputation as 'rainbowfish'.

Many other studies, in particular by David Reznick's research lab at the University of California (Riverside), have shown that plenty of other Guppy attributes, besides colour, can quickly shift and change in the streams of Trinidad: their body size, their lifespan, the age they reach sexual maturity and the number and size of their offspring all respond rapidly to changes in predation.

Reznick's team has measured the rate of Guppy evolution in terms of darwins (in 1949, British scientist J. B. S. Haldane proposed the darwin as a relative unit of evolution).* In one study, Resnick found his Guppies were evolving a rate of between 3,700 to 45,000 darwins. The fastest rates measured in the fossil record are between 0.1

* One darwin is defined as the change in a particular trait by a factor of *e* in one million years, where *e* = 2.718 (or thereabouts).

and 1 darwin. Some say it's meaningless to compare the two time scales, of Guppies changing over a matter of months and ancient species evolving over many millions of years. But Reznick and others argue that the pathways of micro-evolution have a lot to tell us about the coming and going of species throughout the history of life on Earth.

Guppies in the wild are still considered to be a single species, for now at least. In other fish, studies are showing that factors like female colour choice can quickly set up barriers to mating, splitting populations and in time leading to the evolution of new species. And when those colourful barriers break down, species can be lost.

Mix all the colours and get brown

Until 30 years ago, 500 species of cichlid lived in the waters of Lake Victoria, one of Africa's Great Lakes. Since then, however, around 70 per cent of them have gone extinct. The blame is often pinned on Nile Perch. These enormous predatory fish were introduced to the lake in the 1950s as a source of food. They can grow up to two meters (6.5ft) and have a big appetite for cichlids – but not for all of them. The perch ignore at least 200 cichlid species, yet many of them have disappeared nonetheless. There's an additional explanation for the cichlids' demise, one that's only indirectly related to the predatory perch.

Like Guppies, female cichlids tend to prefer mates of certain colours, and it's now widely accepted that this is the only thing keeping many cichlid species from interbreeding. There is a common misconception that distinct species are incapable of mating with each other and producing fertile offspring. Actually, different species can successfully breed together quite often, it's just that they usually don't because other things stop them, either physical barriers like a mountain range, or behaviours, such as females only mating with males of a certain colour. Lake Victoria's cichlids evolved so rapidly, in perhaps as little as 12,500 years, that no other barriers to interbreeding have had a chance to arise – only colour choice. And over the last few decades, that single, vulnerable barrier has been dismantled as the lake's waters have changed.

Back in the 1920s, if you'd gone for a swim in Lake Victoria, ducked your head underwater and opened your eyes you would have been able to see around five to eight metres (16–26ft) in front of you. If you did the same thing in the 1990s you might only have seen a metre (3ft) ahead. The lake's waters have been getting murkier because fertilisers from farmland runoff are stimulating blooms of phytoplankton in the lake. Sediments have also poured in as surrounding forests were felled and roots loosened their grip on the soil, leading to erosion. And the reason the trees were cut down? To build fires for smoking all those introduced Nile Perch.

Studies in the lab and the wild, many led by Ole Seehausen, now at the University of Bern in Switzerland, have examined the effects of Lake Victoria's cloudy water on its cichlid species. In aquarium tanks, when female cichlids are illuminated in monochromatic light (a good approximation for what's happening in Lake Victoria), they can no longer distinguish between blue and red males of different species. In the gloom, the males' colours look washed out and indistinct, in a similar way that red colours are lost the deeper underwater you go. Because the females can no longer tell who is who from their colours, they mate at random and the barrier to interbreeding comes down.

The same thing seems to be happening out in the wild. The parts of Lake Victoria with clearer waters are home to the most cichlid species, whose colours remain bright and distinct. In these areas females can still see the males' colours and pick their own species to mate with. But the murkier the water, the more drab the fish and the fewer species present.*

Cichlids are in peril not simply because they're hunted by introduced Nile Perch but because they're having a hard time seeing each other's colours. Gloomy waters are interfering with cichlid mating habits. Where the lights are dimmed, species are interbreeding, losing their vivid colours and collapsing in on each other. Unwittingly, humans are stealing away species, and at the same time making the world a less colourful place.

* Researchers are also finding more hybrid fish, the results of this interbreeding.

The salmon of knowledge

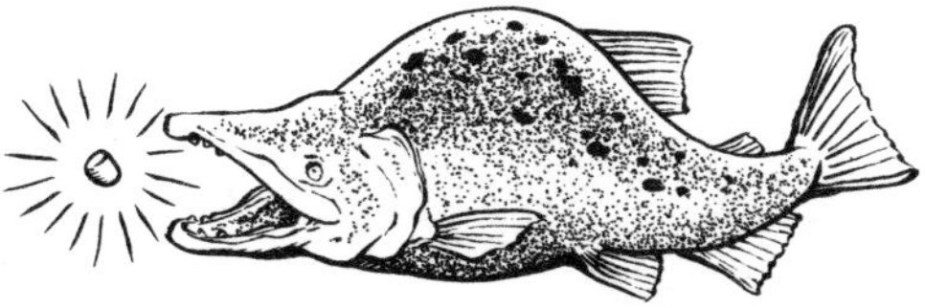

Ireland, traditional

There was once a well with nine magic hazelnut trees growing beside it, and if you peered in you might have glimpsed a large, gleaming salmon circling in the water below. One day, each of the trees dropped a nut into the well, and when the salmon ate them it gained all the knowledge in the world. A prophecy said that the great poet, Finegas, would catch that salmon and eat it, and then he would know all there was to know.

For seven years Finegas sat beside the well with his fishing rod and tried to hook the salmon. When finally he did, he asked his apprentice, Fionn MacCool, to cook it for him. 'But whatever you do,' Finegas warned, 'do not eat the salmon.' Fionn did as he was told and carefully prepared the fish, roasting it over a peat fire. But as he reached in to turn the cooking salmon, a drop of hot oil spattered onto Fionn's hand. With a jolt of pain, the boy stuck his burnt thumb in his mouth to soothe and cool it.

When Finegas saw Fionn, he noticed a new fire shining in his eyes. 'Did you eat the salmon?' he demanded. Fionn pleaded that he hadn't, but he told him how the sizzling hot oil had landed on him. Then Finegas knew what had happened. All the salmon's knowledge had passed to Fionn. From then on, Fionn could call upon all there was to know in the world simply by sucking his thumb, and he became a great warrior and leader of the Fianna, a famous band of roaming Irish heroes.

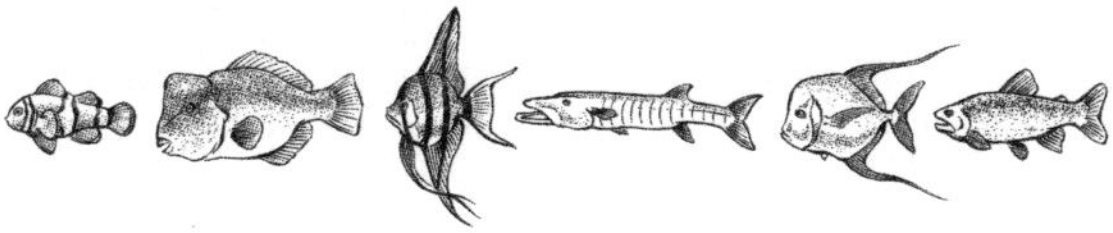

CHAPTER FOUR:
ILLUMINATIONS

CHAPTER FOUR

Illuminations

Before 1815, when Humphry Davy invented his safety lamp, British coal miners sometimes went to work carrying a bucket of dead fish. Naked flames had the dangerous tendency of blowing up any methane that seeped into the tunnels, so miners needed alternative light sources. A bucketful of putrefying fish could apparently produce enough dim, cold light for miners to see by. It wasn't the fish themselves that shone in the dark, but bacteria that settled and began decomposing the flesh and bones, and especially the eyes.

At around the same time that dead fish were replaced by the march of technological innovation, in the early part of the 19th century, scientists were beginning to discover fish that glow while they're still alive. On a three-year whaling voyage in the 1830s, the surgeon and naturalist Frederick Debell Bennett saw 10 glowing fish brought to the surface in a tow-net. Placed in a bucket of seawater on deck, the little *Scopelus* fish swam around and glowed brightly from their scales and rows of dimples long their bodies. When they died, their lights went out.

Many invertebrates glow in the dark: corals, clams and jellyfish, millipedes and centipedes, krill and squid, and of course those famous flashers, the fireflies or lightning bugs, that aren't in fact flies or bugs, but beetles. There are fungi that glow (and no one really knows why) but, as far as we know, there are no glowing plants. Similarly, if you switch off the lights, you won't see any birds, mammals, reptiles, amphibians or any other backboned creatures making their own light. Among the vertebrates, bioluminescence is unique to fish.

It was only when research efforts began to focus on the very deep sea that it became obvious just how many fish glow in the dark, since that's where most of them live. For a long time it was assumed that the ocean depths were empty and lifeless. After all, how could any living thing tolerate such immense pressure and utter darkness? Gradually, however, the idea caught on that it might be worth checking, just in case there was indeed something interesting lurking down there in the dark. And when people started looking into the depths, they uncovered a whole new way that fish are brilliantly adapted to life underwater.

A turning point in deep-sea exploration took place on 7 December 1872, when a former British warship, HMS *Challenger*, set off from the Isle of Sheppey in Kent in southeast England.

The Royal Society in London had enquired of the Royal Navy whether they could borrow a ship for the purpose of studying the oceans on a very long, very ambitious transglobal expedition. The Navy said yes. Guns and ammunition were stripped out and replaced with storage areas and research labs. And with 21 naval officers, 216 crew and six scientists on board, HMS *Challenger* embarked on a very different kind of voyage to her previous missions.

For a thousand days the expedition traced wide loops across the North and South Atlantic, skirted the southern reaches of the Indian Ocean, sailed down the middle of the Pacific and approached the 'Great Southern Ice Barrier', as Antarctica was then known. The journey was almost 70,000 nautical miles long. If the ship could have sailed around the equator, it would have completed more than three whole laps of the planet.

At various points along the way, *Challenger* paused to drop down scientific instruments and measure the oceans' vital statistics in ways never done before. Rope and piano wire were essential items of equipment on board, measuring more than 400km (250 miles) in total. Lead weights were lowered over the side to measure how far below the seabed lay. In the Pacific Ocean, 8,184m of line (26,850ft) were paid out into the Marianas Trench, in a spot later named the Challenger Deep after it was established that the expedition had discovered the deepest point in the oceans.* Ropes were also used to drag trawl nets and dredges through the water, to catch living things from the deep. The *Challenger* scientists probed the 'twilight zone', the region between around 200 and 1,000m (650 to 3,300ft) down where dim, blue sunlight trickles from above and it's not quite dark. They sent nets even deeper, into the permanent dark of the 'midnight zone' below 1,000m (3,300ft), where sunlight doesn't reach. And everywhere they looked, they found hoards of strange creatures. Both at sea and for decades afterwards in laboratories around the world, scientists involved in the *Challenger* expedition revealed a radical new view of the oceans. They discovered there are far more, far stranger things living deeper down than anyone ever imagined.

Fewer than 30 deep-sea fish species were known of before the *Challenger* expedition, and they had only been found in the first hundred fathoms or so of water, less than 200m (650ft) down. On its return to Britain, HMS *Challenger* unloaded 144 species that were new to science, gathered from as far down as 2,900 fathoms – more than five kilometres (three miles).

The fish collection, now housed at the Natural History Museum in London, showed for the first time just how diverse and peculiar deep-sea fish can be. There are

---

* The latest measurements show the Challenger Deep is 10,916m (6.7 miles) deep.

hatchetfish, so named because you could conceivably grab one by the tail and use its metallic, sharp-edged body to chop an imaginary log. Gulper Eels are all mouth; their colossal, hinged jaws will gulp down any prey, big or small, and stash it in a stretchy stomach pouch. There are halosaurs (meaning 'sea lizards') with long, eel-like bodies and flattened, triangular snouts. And there's an array of creatures that look like fleshy, partially deflated footballs, with terrifying jaws and a prong protruding from their foreheads. They all belong to an order called the anglerfish, which includes such things as black seadevils, wolftrap anglers and warty seadevils.

British marine biologist, John Murray, wrote about all these fish in part of the immense, 50-volume report on the expedition's findings. He had been on board *Challenger*'s four-year voyage, and had seen that many of the freshly caught fish shared a remarkable attribute: they twinkled with constellations of glowing dots, or oozed shiny slime across the ship's deck.

Back then, scientists had no way of venturing into the deep sea to watch living fish and witness how those luminescent body parts and seeping goo were put to use. All they had to work on were the fish's bodies. Luckily, though, even after being hauled up from the deep, many of the fish were in remarkably good shape, and gave hints about what they got up to down in the depths.

Lanternfish, Murray thought, might have darted through the darkness, flashing lights on their tails to lure prey within reach before quickly turning around to chomp them. He imagined a similar scenario for the Obese Dragonfish. The length of your arm, it has a mouthful of fangs and, dangling from its chin, a long barbel with a glowing tip. This glowing spot, Murray was convinced, must be a lure to attract other fish into the dragonfish's reach. Its jet-black skin is speckled with white glowing dots, which Murray envisaged might startle approaching predators with bright moving patterns that looked, as he put it, like 'shadows of clouds'.

Many fish in the *Challenger* collection have glowing sacs near their eyes that could feasibly have made light bright enough for the fish themselves to see by. They could 'shoot rays of light in a direction in which they want to explore', Murray wrote. When they wanted to stay hidden, these fish could draw down shutters and blink their headlights out.

Now, almost 150 years later, scientists are still making discoveries in the deep sea. A few lucky people get to dive many miles beneath the waves inside submersible vehicles, where they can peer through thick acrylic windows at the fish glowing and flashing outside. Fleets of underwater robots, known as Remotely Operated Vehicles or ROVs, are sent down with cameras that beam real-time images to the surface, and some glowing animals are brought up alive. Studies like these are finding that John Murray was right in many of his theories about how animals use bioluminescence: flashlight fish do seem to use the lights under their eyes to see in the dark; anglerfish do use their glowing barbels and beards to lure in prey. Research teams are also finding there are plenty of other things going on in the fish's dark realm.

Welcome to the blue-light district

The latest census lists 1,510 ray-finned fish plus 51 shark species that make their own light. Bioluminescence didn't just evolve the one time in fish; it repeatedly evolved among distantly related species on at least 30 separate occasions. Fish don't simply glow in the dark, they are the uncontested champions of bioluminescence.

There are two main ways they do it. For just over half of them, an intrinsic glow comes from a chemical reaction in their bodies. They've evolved one or more genes for enzymes known generically as luciferases, which include a whole range of different molecules. These enzymes speed up reactions with a light-emitting molecule known as luciferin. Reacting with oxygen, bonds break within these molecules, shooting out photons of light.

Precisely where the light-making molecules come from is not well known. All sorts of glowing marine creatures use the same four types of luciferin molecules. Fish mainly use two of them (called cypridina and coelenterazine) and the same ones are used by bioluminescent squid, shrimp, jellyfish, brittlestars and plankton. It might at first seem a little odd that so many different animals use the same light-making molecules, but they probably derive them from their food – and many of them eat similar things. This certainly seems to be the case for one glowing fish, the Midshipman Toadfish. Across their range, along the Pacific coast of North America, some of them glow and some don't. In the south, members of a Californian population are studded with hundreds of brightly glowing dots, called photophores. But further north in Washington's Puget Sound, toadfish have those same structures in their skin but they don't light up. However, if you feed them the right food, those northerly fish begin to illuminate. They need to eat a certain type of minute crustacean called copepods, one that only lives in the south. Presumably the copepods provide the Midshipmen's luciferin, and without them the fish up north remain in the dark.

The other half of the bioluminescent fish species don't have an inherent ability to glow but become radiant thanks to hordes of bacteria living inside them. These glowing microbes are widespread in the oceans. It's quite easy to grow colonies of them: just swab a piece of flotsam that's washed up on a beach, wipe it on a Petri dish, incubate it for a while, and among the microbes that multiply and creep across the dish there will be some that glow. Alternatively, if you can bear the smell, find a dead fish and wait for it to start glowing, as the coal miners once did.

Superabundant bacteria are probably responsible for a rare and eerie phenomenon when the sea glows with a constant, dim light.* Occasionally conditions lead to slicks of free-living bacteria that are dense enough to trigger each other to glow. One of these so-called milky seas formed in 1995 off the coast of Somalia. It was captured on a satellite image that showed it covering some 15,000km^2 square kilometres (6,000 square miles); it was composed of an estimated 40 billion trillion bacteria.

A long-standing theory holds that oceanic bacteria evolved their glowing tendencies to *encourage* fish to eat them. Bacteria colonise clumps of organic sea-going matter, things like fish faeces and moulted crab and shrimp shells. Making these scraps glow increases the chance that fish will spot and eat them, thus delivering the bacteria into their ideal living quarters – the interior of a fish's guts. It's then a logical step for many fish to positively encourage such invasions and begin using borrowed light.

Fish have evolved special organs that house their bacterial associates, including the dangling fishing rods on anglerfish foreheads, structures formally known as esca. And while most fish either make their own light or use on-board bacteria, there's at least one species that does both. The Illuminated Netdevil, a type of anglerfish, has a bulbous lure sprouting from its head, which glows with bacterial light and looks like a small pickled onion; in addition, it has a beard hanging down from its chin like a frond of seaweed, which glows with light from its own internal chemical reactions.

The fact that so many fish have evolved bioluminescence, so many times, indicates just how useful it can be. Living in

---

* This is different from the flashing twinkles in tropical seas, stirred up by waves, boats and playful dolphins; those bioluminescent lights come from disturbed dinoflagellates, members of the plankton.

the dark and being able to make and control their own light gives them an immense advantage over everything else.

A lot of fish use their lights not to be seen but to become invisible. Inhabitants of the twilight zone, 1,000m (3,300ft) down, run the risk of being spotted by predators looking up and seeing a dark silhouette cast against the blue ceiling above. A similar thing happens if you step outside on a clear night shortly after sunset, and look up at the darkening sky to see a bird or a bat fly overhead: you notice a dark shape flapping across the indigo sky. To counter this, some fish have dots of light arranged in distinct lines that break up their outline and make them difficult to recognise (in a similar way to the Emperor Angelfish's blue and yellow stripes). And many have bellies covered in photophores that cloak them in blue light, disguising the whole of their dark shadow – a phenomenon known as counter-illumination. They can even adjust their belly lights to precisely match the intensity of blue light seeping down from above, making sure they stay hidden at different times and depths. It's a tactic employed by a host of twilight-zone fish, including bristlemouths and lanternfish, those most numerous vertebrates on the planet. They spend most of their day in this shadowy realm, doing a good job of pretending they're not there at all.

Similarly, Cookie Cutter Sharks have blue glowing bellies. Frederick Debell Bennett caught these half-metre (20in), spindle-shaped sharks on his 1830s whaling voyage and described their light as giving them 'a truly ghastly and terrific appearance.' He also noted around their necks a dark, non-glowing band, tapered at both ends. This, he thought, might look like the shadow of a smaller fish. In this way, Bennett proposed, Cookie Cutters might lure in larger, faster animals – dolphins, whales and tuna. When the hunters get close, expecting to find food, they instead lose a chunk of flesh to the Cookie Cutter. The parasitic shark latches onto their skin, rotates on the spot and takes a big mouthful, before swimming off and leaving behind a characteristic, circular wound* (hence their name). If this theory is correct, then it would explain how slower-swimming sharks catch up and bite the fast-moving predators, which are commonly found pocked with cookie scars. And it would also make Cookie Cutters the only animals known to use an absence of light to mimic other species, like shadow puppets.

Other fish use their blue lights to talk to each other. Like a firefly, if you flash a light at a lanternfish it will blink back, perhaps mistaking you for a potential mate. Barreleyes, also known as spookfish, write messages to each other on their bellies. Little was known about these fish for a long time because the delicate specimens always got mangled in the nets bringing them back up to the surface. Then, in 2004, a living Pacific Barreleye was filmed 600m (2,000ft) underwater, off the coast of California. It has enormous green, tubular eyes that swivel around beneath a clear bubble like an astronaut's helmet; this probably protects their eyes while they're grabbing bits of food stuck to the stinging tentacles of siphonophores, relatives of jellyfish.† Those telescopic eyes usually gaze upwards to watch for silhouettes of animals swimming above them. Barreleyes also have offshoots from each tube that act as a second, smaller pair of eyes. Instead of the standard clear lens, these have a shiny layer of guanine crystals that focuses light – they're the only animals that have mirrors for eyes – and they gather the dim blue glow of bioluminescent animals, including other barreleyes. They have a structure towards

* Unlike most other sharks, which replace their teeth one at a time, Cookie Cutters' razor-sharp teeth are neatly interconnected, meaning they have to lose theirs all at once, like spitting out a set of dentures.

† The famous Portuguese Man o'War is an example of a siphonophore.

the end of their intestines, called a rectal bulb, which is infested with glowing bacteria. The light is channelled across their belly, which is flat like the sole of a shoe and lights up like a bicycle reflector. This not only hides the barreleye's silhouette but different species have particular patterns of dark pigments across their glowing soles. In this way, barreleyes broadcast their identity into the dark, perhaps so that other members of the same species can spot them.

Closer to the surface, another group of glowing fish use their lights to communicate at night and in murky waters. Ponyfish are small and silvery and generally hang about in shoals in estuaries and coastal waters. Catch them at the right time and you'll see a whole shoal flashing in synchrony. Male ponyfish are the chief flashers. They have a ring of glowing, bacteria-rich tissue looped around their oesophagus, and a silvery swim bladder that reflects the light. Transparent windows in the side of their bodies let the light shine through and shutters draw down when they want to block them off. Without their lights on, it's difficult to tell different ponyfish apart. Once they illuminate, though, each species shows particular constellations and sequences of flashes. Females do glow, but they generally aren't as bright as their male counterparts (their glowing throat rings are up to a hundred times smaller than the males'). Male ponyfish are talking to females with their lights, tempting them to come and see what all the flashing fuss is about.

By helping fish recognise each other – rather like colourful patterns in Guppies and cichlids – it seems bioluminescence could play an important part in the evolution of species, especially in open waters where there are no physical barriers to split populations. Fish that use their lights to communicate are often the groups with the most species. There are 252 species of lanternfish and they all have their own luminescent patterns, on their heads and tails and down their sides; their shining motifs are beacons of their identity. Meanwhile, bristlemouths only use their lights to disguise their silhouette and not to send out messages into the dark. In total there are only 21 known species of bristlemouths. Studies have found that the rate at which new species evolve is higher among deep-sea fish that use light to attract mates and recognise each other, as lanternfish do. The same goes for deep-sea sharks. Lanternsharks are among the smallest of all the sharks – many of them are roughly a human handspan from head to tail – and there are 38 described species, a good number for a shark genus. They inhabit the twilight zone and light themselves up in various ways. Velvet-belly Lanternsharks have sharp, toxic spines that illuminate like light sabres, presumably to warn off predators. They also have glowing patterns along their bodies, on their flanks and tails. Captive sharks often rotate their bodies left and right while they're swimming, so that spectators see a flashing signal. The males even have glowing claspers – the elasmobranch version of a penis – that intermittently flash on and off, presumably putting on an irresistible, dazzling display for the female sharks.

So far, all of these fish-made lights have been a similar colour – most are blue, to match both the downwelling light in the twilight zone and the visual pigments in many open-ocean fish, which generally see best in blue. But there's one group of glowing fish that doesn't conform to this blue uniform.

Dragonfish break all the rules and make red light. They shine beams of it through the dark, hunting for prey and communicating with each other on their own private wavelength, like night-vision goggles. Making red light lets dragonfish foil the camouflage of red-coloured creatures. Red pigments, which are normally rendered black by the lack of red light, suddenly stand out in the dark. The dragonfish not only make red light, but they see it, too. And they've adjusted their vision in a most remarkable way.

One dragonfish species, the Stoplight Loosejaw, has bulbous red eyes. Their retina detects red light using an unusual pigment that acts as a photo-sensitiser – it boosts the fish's vision in the far-red part of the spectrum. It's a modified version of chlorophyll, a molecule that plants, algae and bacteria use to harness energy from sunlight. As far as we know, animals can't make chlorophyll, so the Stoplights must be getting it from their food. But exactly how is something of a mystery.

This particular chlorophyll molecule is made by a type of bacteria that live in mud near the shallow edges of the sea. The microbes have never been found in deep, open water, where Stoplights live, and yet somehow this chlorophyll gets inside copepods (those tiny crustaceans again) on which these fish feed. How the link is made between the shallows and the deep is not yet known. It's a strange thing indeed that fish living hundreds of metres beneath the waves have borrowed a pigment that usually catches sunlight, and co-opted it for seeing in the dark.

Secret graffiti

While bioluminescent fish were known of back in the 19th century, it would be many decades before the discovery of another major group of light-emitting animals. In 1927, British naturalist Charles E.S. Phillips wrote a short letter to the journal *Nature* describing the glowing anemones he had seen clinging to a rock at the beach in Torbay, on Britain's south coast. He took a few of these flower-like animals with him back to London, shone ultraviolet light on them, and saw their tentacle tips shining bright green. Phillips suggested in his letter that UV lamps would be a useful addition to a marine biologist's research tools, but no one at the time followed his lead.

It was another 30 years before anyone tried taking UV lamps underwater. In the late 1950s, Richard Woodbridge dived with his homemade lamps in the chilly waters off Maine, in the northeast United States, and saw invertebrates lighting up around him. Like Phillips, he also wrote to *Nature* highlighting this useful new research tool, but again it didn't stir much interest. Woodbridge lent his lights to science-fiction writer and keen diver Arthur C. Clarke, who tried them out, then featured them in his 1963 book *Dolphin Island*. Clarke sent his protagonists on a dive with UV lights on Australia's Great Barrier Reef. 'When this fell upon many varieties of corals and shells,' he wrote, 'they seemed to burst into life … blazing with fluorescent blues and golds and greens in the darkness.'

Eventually biologists caught on and started shining UV lights at all sorts of animals, finding that many glow back. Spiders, scorpions, budgerigars, butterflies, zooplankton, corals, molluscs and mantis shrimp all put on these glowing displays. They do this not by making their own light but by manipulating the light that's already around them. This is not bioluminescence, it's fluorescence. And researchers began to find that lots of fish were covered in this form of secret graffiti.

On the eastern side of Drawaqa Island, one of a chain of forested islands in Fiji, lies a stretch of sand known as Sunrise Beach. Night has just fallen, and when I see the full moon reflecting on the sea I decide it should be renamed Moonrise Beach.

After a short trek across the island from the dive shack, my scuba gear is digging into my back and I enjoy buoyant relief as I wade into the cool water. For a few moments I float there, pulling on my fins and sorting through more bits of kit than I normally bring diving.

I test the two waterproof lights that dangle on cords from my wrists; one shines regular, white light, the other is deep blue. Around my neck I have a plastic yellow visor that, when the moment comes, I will flip over the front of my dive mask. It makes me feel like a Lego space-woman.

Then I push a button and let out air from my dive jacket; I drift down into the black water, and feel a sense of serenity fold in around me. I've always loved night diving. The first time I did it I expected to be cold and scared and lost in the dark. I thought it would feel as if I was creeping into a forest at night, my mind inventing unseen beasts that skulked just beyond the reach of my torch beam. For some reason being underwater at night is completely different. It feels tranquil and meditative, with the comforting dark water and peaceful, sleeping fish.

Ahead of me my dive buddies are illuminated in their own pools of light, and their exhaled bubbles trail above them like silvery prayers rising to the sky. My white torchlight interrogates the darkness, putting back the reds and oranges that are so quickly lost from sunlight seeping down from above. Everything looks more vivid than when I dived in this same spot earlier on, with just daylight to see by. Gently I sink and kneel on the sandy seabed, and settle in for a dive like I've never done before.

With my yellow visor in place I switch off my white light, and for a thrilling moment sit in utter blackness. I close and open my eyes and it makes no difference. Then I push the button on my blue light, and instantly the world around me transforms.

Seconds ago the reef was dominated by subdued greens and browns. Now it's become a strange glowing wonderland. The backdrop is deep, velvety crimson. Branching corals reach up with neon green fingers tipped in purple. Brain corals are covered in green and red meandering valleys. Scattered everywhere across the reef are dots of light as if the starlit night sky has fallen all the way into the sea.

Blue light is my magic wand. Wherever I point, the reef and its inhabitants glow back at me. A tiny sea snail glides along on a pea-green foot, towing a bright scarlet spiral of a shell. A solitary anemone the size of my outstretched hand undulates its glowing yellow tentacles through the water, drawing them one by one into the middle as though licking its fingers.

Just to check I'm not dreaming, I flick my white light back on and briefly restore normality to the reef. Then I go back to the blue light and tumble once again down the rabbit hole. In front of me sits a lizardfish. Normally, these aren't the most exciting species. They have mottled, beige coloration to blend in with the seabed where they sit still, hoping not to be seen. Under the beam of my blue torch, this one is now shining from head to tail in a lurid acid green, which casts a bright shadow on the sand around it. And there's a goatfish with two long whiskers under its chin, which look as if it's been dipped in fluorescent yellow paint, and a grouper resting on the seabed with a mottled red glow.

Not all the fish I see are shining bright colours in my blue light. A Moorish Idol hunches in a small cave, indistinct and grey, like a poorly developed negative of itself. But when I peer under a ledge of coral I spy a small fish underneath, a bream, with its back to me. In daylight this species is half white, half black with white lines running back from its eyes. Now, at night and bathed in my blue light, this one has new green stripes across its flanks. I watch for a few moments until it turns around and pouts at me with lips painted in bright, glowing red lipstick.

Until that dive, I'd never noticed the fluorescent patterns on fish. They've always been there, but I missed this kaleidoscopic world because, like most other scientists and divers, I hadn't been looking at them in the right way.

All the glowing animals I see in Fiji have fluorescent pigments in their skin that play with colours and create wavelengths that aren't normally there. The pigments absorb the blue light in my torch beam, then re-emit light of a different colour. Usually, shorter wavelengths of light – ultraviolet or blue – get absorbed and re-emitted as longer wavelengths, pushing the colour up the rainbow, to greens, yellows and reds. This happens when photons of light excite electrons in the pigment, sending them briefly into a higher energy state before they relax back down and release the energy again. It's a swift transition and means fluorescent molecules only glow while light is still shining on them, unlike phosphorescent watch dials or glow-in-the-dark stars on a bedroom ceiling, which soak up photons and re-emit light for a long while afterwards.

Various fluorescent substances interfere with the wavelengths of light that fall on them. A common fluorescent molecule is chlorophyll. This is why many corals glow under blue light: the single-celled algae living inside their tissues contain fluorescent chlorophyll, which shifts blue light into red.

Probably the most famous glowing sea creature, at least in scientific circles, is a jellyfish that happens to be both bioluminescent and fluorescent at the same time. In the wild these small, delicate creatures, known as Crystal Jellies, pulse through Pacific currents along America's west coast. They're transparent but when they bump into things they flash green. This light show is a two-step process. First, a chemical reaction in a protein called aequorin produces photons of blue light – the bioluminescent part of the proceedings. This blue light then shines on another protein, a fluorescent one, which shifts the wavelength and makes the jellyfish glow green.

This second molecule, known as Green Fluorescent Protein or GFP, was first extracted from these jellyfish in the 1960s and has gone on to revolutionise scientific research.* A cloned version of GFP is now ubiquitous in laboratories,

---

*Various other fluorescent proteins have since been discovered in other animals, but the jellyfish version remains the most widely used.

where it allows researchers to tag specific genes and see where and when they switch on inside a living cell or parts of a body, simply by illuminating them in UV or blue light. GFPs have shone light on spreading cancer cells, and tracked nerves as they grow and connect up inside brains. They've even been used to genetically engineer fish that glow in the dark, originally as a test for pollutants. When a genetically modified Zebrafish swims through contaminated water it will let you know by shining brightly. Glowing GM fish are also now sold as pets. And in just the last few years it's emerged that lots of fish are naturally fluorescent without the need for genetic tinkering. Their secret graffiti was only uncovered following an accidental discovery.

It was on the shores of the Red Sea in Egypt that Nico Michiels, a marine scientist from the University of Tübingen in Germany, decided to go diving with his mask covered in a red plastic film. He wanted to see for himself just how quickly red light is lost from sunlight as he descended into the sea.

'It felt a bit creepy,' Nico tells me, as we chat over Skype. At five metres (16ft) down it was already getting dark. By ten metres (32ft) he felt as if he was diving at night, even though it was midday in the tropics. All the red light in the water around him had gone and with his modified mask blocking out other wavelengths there was no other light for him to see by.

'I couldn't really see my dive computer,' he says. 'I could hardly see my dive buddy.'

Gradually, Nico's eyes adapted to the gloom and he began to notice corals on the reef dimly glowing red; this was the chlorophyll of the coral's symbiotic algae emitting fluorescent red light.

Then he suddenly spotted a pair of tiny red eyes shining at him. A similar pair of eyes stares at me from my computer screen. Nico's profile picture on Skype is a goby that looks as if it's wearing a pair of big red spectacles.

'I got totally excited about this,' he tells me. On that dive, all the other fish were almost invisible to him except for the ones with piercing eyes. He knew then, immersed in this strange red world, that he had seen something special.

At that point, there were no published scientific papers on fluorescent fish. Nico thinks one reason the glowing fish were overlooked for so long was that people were focusing on night dives, using UV lights, when most fish are asleep and tucked away out of sight. I'd been lucky to spy a few fish on my nocturnal foray in Fiji.

'Let's be honest, nobody had ever been stupid enough to put a red filter on their mask,' Nico tells me, 'because it's quite predictable – you're not going to see anything.'

But what Nico did see inspired him to embark on a whole new area of research. Until then he'd worked mainly on the sex lives of earthworms and flatworms. But following that red dive he turned his focus on glowing fish. He spent the next few years searching for them and diving all over the world with his red mask on. Nico and his research team brought fish back to his lab in Germany and, shining blue and UV lights on them, showed that many are not bioluminescent but fluorescent.

In the first paper on the subject, in 2008, Nico's team identified more than 30 fluorescent fish. A lot have red rings around each eye; some glow red all over. Red isn't the only colour that fish glow. Following Nico's initial discoveries, John Sparks from the American Museum of Natural History led a team that collected coral-reef fish from various oceans, and bought others from the aquarium trade. Using blue lights, they saw that as well as glowing red, some fish glow green, some glow orange, and some have multi-coloured patterns and look as if they're living in a permanent DayGlo disco. Not only that, but these fish are stationed all across the evolutionary tree; there are fluorescent sharks and stingrays, flatfish and stonefish, blennies and gobies, triggerfish, seahorses, eels, mullets and wrasse.

By the time Sparks' study came out in 2014, Nico Michiels was utterly convinced that plenty of fish are fluorescent and he set about a bigger task: he wanted to know why this phenomenon evolved, and what the fish use their fluorescence for. As he saw on his red-masked dives, most of the sun's red light is absorbed and gone within the top 10m (33ft) of water. Without red light shining on them, red pigments lose their colour and appear grey or black (which, as we saw earlier, is why red is generally a good colour to use as camouflage in the deep sea). Fluorescent sea creatures subvert this rule of physics by converting available blue light into absent red. They're making colours that are normally missing.

It's important to realise that pointing a blue light at a fish and seeing it glowing back, as I did in Fiji, is not exactly natural. It shows us that there are fluorescent pigments in their skin, but this is not how fish see each other. Unlike the Crystal Jellies laden with bioluminescent aequorin, most shallow fish don't carry around their own blue lamps to shine on themselves. Fluorescence is naturally a low-key affair.

I did, though, do one thing the same way as the fish, by wearing a yellow cover to my dive mask; many fish have yellow eyeballs, and see the world as if they were wearing yellow-tinted sunglasses, which accentuates longer wavelengths of light, the reds, oranges and yellows, and potentially boosts their perception of fluorescent colours. And that could be important because when there isn't a human diver with a torch, fish fluorescence is much more subtle. On that night dive in Fiji, it's more likely that moonlight could have naturally lit up the fluorescent fish around me. And during the day, it's the abundant blue part of sunlight in the water that brings out the fish's fluorescent colours.

The big question is why do they produce these pigments, and especially why has fluorescence evolved so many times among the fish? In their latest survey, Nico's team found 272 red-fluorescent fish species. A lot of them are hunters that spend their time hunkered on the seabed hoping no one will notice them. These ambush predators include scorpionfish and flatfish, which rely on camouflage to trick unsuspecting prey into wandering within striking range. Fluorescence makes them even harder to spot, covering their skin in sprinkles of patchy fluorescence that match the weedy, chlorophyll-rich backgrounds of a reef.

Just like many of their bioluminescent cousins, some fish seem to use their fluorescence to communicate. A lot of them have fluorescent patches on their fins, which they can flash at each other, then quickly fold away before a predator spots them. Male Fairy Wrasse have red fluorescent patterns on their faces that may help them recognise intruders. These are angry little fish; presented with a mirror, they strike a threatening pose and try to pick a fight with their own reflection, mistaking it for another fish. At least they do when illuminated in regular, white light. When Nico's colleague, Tobias Gerlach, put filters in front of the mirror that blocked out the red wavelengths, the males couldn't see the fluorescent facial patterns and they became much less aggressive. This suggests that fluorescence could be poster colours that shine in a way Konrad Lorenz might never have expected.

Nico has one more idea that could help explain why so many fish are fluorescent. The latest study, headed up by Nils Anthes, showed that red fluorescence is especially common among small predators such as gobies. They hunt for even tinier prey, things like shrimp that are well camouflaged on the seabed and extremely difficult to spot, but there's one thing that could give them away: their eyes. Nico and his team think the fluorescent eyes of these little fish, like the first one he spotted in the Red Sea, may shine brightly enough that at very close range the eyes of their prey shine back at them.

Take a flash photograph at night of a cat or a crocodile, and you'll see their eyes reflecting out of the dark. Fluorescent fish could be doing something similar, only using their own eyes instead of a flash bulb. The red light from one pair of eyes might make another pair shine. The shrimp, Nico says, 'are very well camouflaged, but their eyes are not. If you can make their eyes glint, you break their camouflage.'

His eye-gleam theory could mean that some fluorescent fish are using their light-bending trick in a similar way to the bioluminescent headlights of flashlight fish and Stoplight Loosejaws, only with much less dazzling displays. As Nico points out, not all fish want to be so obvious and stand out. 'They are bound to have very subtle mechanisms.'

Being relatively new to the field of fish vision, however, Nico is having to work hard to get his ideas accepted by more established researchers. 'It's going to take quite some time to convince everyone else,' he says. His team are looking at fish fluorescence from every conceivable angle: what the fish's eyes detect, how it affects their behaviour, how they control the fluorescent patterns across their bodies. 'We're now just building up our arguments,' he says.

While they do that, Nico will keep on trying to understand what's going on behind the eyes of those fish that he accidentally saw gleaming at him while he wore his red mask. As he says, 'There's more hints on the way.'

O-namazu

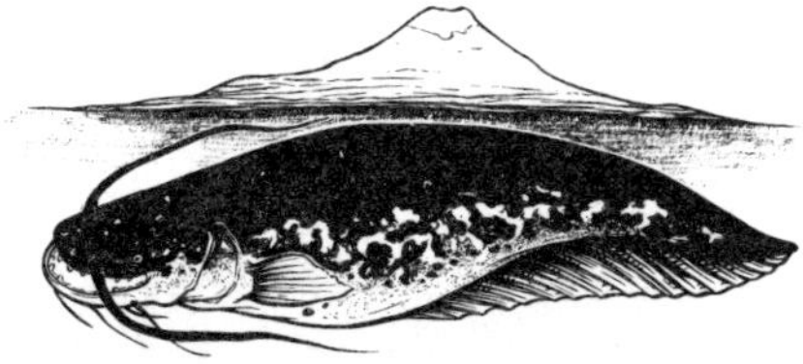

Japan, Edo period

Hidden beneath the islands of Japan lies a giant catfish that people had known of for a long time. They called him O-namazu. The only thing stopping O-namazu from bringing great disasters and misfortune to the world was the deity, Takemikazuchi, who held the catfish down with a large stone. Then one day Takemikazuchi went away to meet with other gods in a secret temple, and he left Ebisu, the god of fishermen, in charge of guarding the great catfish. But Ebisu got drunk and dozed off, and O-namazu wriggled free. The catfish thrashed his tail and unleashed a terrible earthquake. It destroyed much of the city of Edo and killed thousands of people.

Those who survived told many stories of the earthquake and the catfish. Some said O-namazu was jealous of the other fish that are more celebrated in Japanese cuisine than the catfish. Others said he was punishing human greed, and forcing rich people to share out their great wealth. And some said that O-namazu didn't want to make earthquakes at all. There were devious men – traders and carpenters – who coerced the great catfish into shaking and breaking the world, so they could make a handsome profit clearing up the disaster and rebuilding the city.

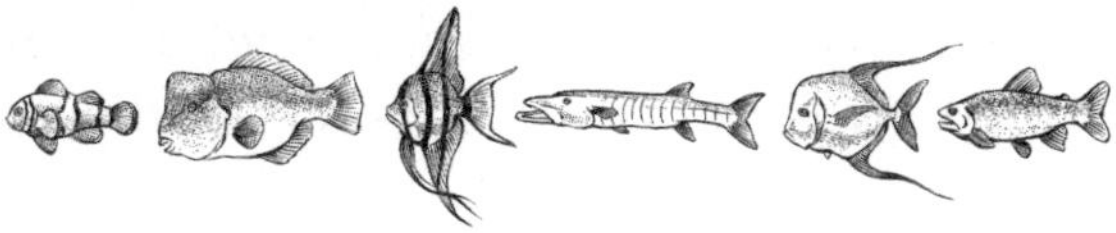

CHAPTER FIVE:
ANATOMY OF A SHOAL

CHAPTER FIVE

Anatomy of a shoal

A sardine swims through the cold waters of the eastern Pacific, and it's not alone. Like many fish, sardines don't do well on their own and always swim with others. This one races along, surrounded by a swirling school of a thousand animals that seem to think as one, turning together, speeding up, slowing down. But the little fish isn't a mindless part of a machine; it's watching and thinking, feeling, listening and deciding what to do next. It somehow knows the unspoken rules that keep the school together.

Choosing which school to join is partly a matter of finding other fish that are a good size match. Rule one: don't be the biggest or the smallest fish that stands out, or you'll be the one a predator spies and eats first. How a fish sizes itself up against its schoolmates is not entirely clear, but they somehow figure out who is bigger and who is smaller.

The next rules ensure the sardine doesn't bump into or get too far from the other fish. Rule two: if the fish behind gets too close – within two body lengths – then speed up. Rule three: if the fish in front gets closer than that, slow down.

In perfect synchrony the school seems to turn all at once, but the fish are not all equal; there are leaders and there are followers. Our particular fish is a follower. Its preferred spot isn't at the front where the leaders swim, but further back in the midst of the pack. It decides which way to turn by watching its neighbours, perhaps just the closest few or all of the sardines in its field of vision. Pressure-sensitive pores along the sardine's body (called the lateral line) let it feel the position of its nearest companions by sensing wrinkles they leave behind in the water as they swim through it.

Suddenly, a wave of panic rips through the school and the sardines all dive down and huddle together. The fish further back haven't seen the sea lion but they get the message of imminent danger, written on the shining, turning bodies of the other fish around them. Like a Mexican wave travelling around a stadium, waves ripple through the school from fish to fish. The waves move much faster than the fish themselves and rapidly transmit vital survival information to the whole school.

As anxiety levels rise among the sardines, they begin to pay even closer attention to what's going on around them and more precisely copy their neighbours' movements. Now they're under attack, it's more important than ever to blend in and do exactly the same as everybody else. Any fish behaving differently could catch the sea lion's eye and become its next target. The sardines blend together until it's difficult to make out a single fish, and they all get lost in the crowd.

The sea lion launches another attack, this time splitting the school in two. The divided fish know they must stay together and like a cascading fountain they flow back to reform a single throng. The hunter hasn't yet made a successful strike, but it has driven the sardines towards the shore, restricting their movements in a sandy cove. Again and again it swoops in, but the sardines seem to predict its every move. It's as if the school can read the sea lion's mind but in fact they're just incredibly fast. Giant nerves transmit signals between the fish's brains and muscles, and they respond within a fraction of a second.

The attacker hardens its resolve and pushes once more through the school. Tension mounts, and the sardines swim faster. The school turns in on itself and forms a tight, spinning sphere. Every fish tries desperately to get to the middle. Each wants to hide behind another and get as far as possible from the hunter's snapping jaws. This selfish geometry shows the fish aren't looking out for one other. They're each just using the school to try and stay alive.

Finally, the sea lion picks off a sardine, then another. Those were the unlucky ones. Inside the school there is still safety in numbers. By sticking together, most of the sardines have escaped unharmed, a far greater proportion than if they were all solitary and roaming the oceans on their own, just lonely pairs of eyes watching for danger.

A life spent in water is not enough to distinguish fish from non-fish, despite the claims in ancient fish books. And yet the ways that fish move through their three liquid dimensions is a crucial part of being a fish.

Back in California, as my 15 year-old self watched those schooling sardines expertly avoiding the sea lion's jaws, I was gazing into a world very different to my own. Fish inhabit a medium that's 900 times denser and 80 times gloopier than air, factors that dominate everything they do. They have to push their bodies forwards to overcome the drag of water that tries to hold them back. On the plus side, though, all it takes is a gas-filled balloon, the swim bladder, and fish can shrug off gravity's pull and float effortlessly in their buoyant surroundings. No bird, bat or insect can fly with such ease.

Other aquatic animals have their own ways of getting around underwater: squid and octopus squirt water jets to propel themselves along; some have flapping 'ears' on the sides of their heads; crabs and shrimp have flattened leg paddles; tinier creatures row through the water with flicks of antennae and hairs. None swim as fast, as furiously or as far as fish. Fish have evolved over hundreds of millions of years to artfully and efficiently swim through the life aquatic.

Different strokes

You can look at a fish and from its shape alone know a lot about how it moves. In Senegal in West Africa, I recently saw a Yellowfin Tuna for the first time. It was propped upright on a bed of crushed ice and staring at the ceiling in a market, and it was massive. If I'd hugged it I would have wrapped my arms maybe half way around its silver body. This fish was all about power and speed. Its torpedo-shaped torso was solid muscle. The tail had been cut off and sat next to it, as if another tuna was diving through the counter; its shape is an elegant crescent, not so good for steering but ideal for cutting down drag while the tuna is cruising. To reduce drag further, during long swims, the pair of pectoral fins would have retracted into slots on each side of the tuna's body, to give it a smooth streamlined profile. When they hunt, the pectoral flips back out to steer and chase after prey. Two elongated fins, also bright yellow and shaped into curving sickles, would have helped prevent the tuna from rolling over as it swam. Rows of triangular, yellow spikes along its flanks probably channelled a streamlined flow towards the tail, helping it shove water sideways and backwards, creating forward thrust.

The man behind the fish counter tried picking up the yellowfin's tail, but struggled with its weight and slippery skin and dropped it on the floor. He tried and failed to pick it up again until someone came and helped him. I wondered how many people it had taken to catch it and bring the fish on deck in the Senegalese fishing fleet out in the Atlantic.

Fish shaped like tuna, with a torpedo body and a forked or a crescent-shaped tail, have evolved for long-distance, endurance swimming: mackerel and swordfish, marlin and sailfish. Swordfish and sailfish have a reputation for being the fastest fish, able to sprint at 100km per hour (over 60mph), but recent studies suggest that this is an inflation of reality. Even so, these hunters are far from lazy. Sailfish can probably manage bursts of almost 32km per hour (20mph), far speedier than any of the smaller prey fish which they chase after, and that is surely the point. If fish swam any faster they'd begin to run the risk of damaging themselves with cavitation bubbles. Fluids under high pressure form air bubbles that then collapse, creating an intense shockwave. Pistol shrimp on coral reefs make cavitation bubbles when they snap their claws (this is where a reef's crackling soundtrack largely comes from). The shrimp's tough carapace can withstand the shocks; fish skin and scales probably can't.

Similar to tuna and sailfish, there are torpedo-shaped sharks with forked tails, like makos and Salmon Sharks, which also swim long distances. They don't have swim bladders and even with oily, buoyant livers sharks do have a tendency to sink. Helping to make up for this, their large pectoral fins are shaped in cross section like aeroplane wings. As they move forwards, water travels faster over the top of their fins than underneath, creating an upward pressure. A 2016 study found that Great Hammerhead Sharks spend up to 90 per cent of their time swimming rolled over on their side, at angles of between 50° and 75° to the vertical, an awkward-looking pose, but one which boosts the lift they get from their tall dorsal fin.

In contrast to long-distance marathon swimmers, fish with wide, fan-shaped tails tend to be fast-start sprinters. Ambush predators, like pike, barracuda and groupers, have wide tails that push aside a lot of water. Big tails are hard work, with a lot of drag, but they're effective over short distances when sudden speed and the element of surprise are all-important.

Eels swim with undulating waves that run along their entire, bendy bodies, usually from head to tail; by switching to waves in the other direction they can swim backwards. Knifefish hold their bodies stiff and swim with undulations of a long fin that runs along their underside; Bowfins do a similar thing, but with a dorsal fin along their backs.

Flatfish move in the same way as regular, upright fish, they simply do it lying on their sides. Plaice, sole and various others spend their first few weeks as hatchlings swimming in a typical, vertical fashion. Then the bones in their skull begin to bend and shift, their mouth changes shape and one eye moves across their face to join the other (either the left or the right eye moves, depending on the species). One side of their flat body becomes pale and white, the other dark and speckled. When they're finally ready, the mature fish lean over and adopt a sideways stance, laying their pale side against the seabed, their dark camouflaged sides uppermost, and look skywards with two eyes on the same side of their head. Now, instead of sweeping their tails from side to side, they undulate up and down. Some elasmobranchs have also adopted this flattened lifestyle, to sit and wait for prey on the seabed, but they do it in a different way: skates and rays squash themselves from top to bottom, press their bellies to the seabed and swim by flapping large pectoral fins stretched out to the sides like wings.

Flying fish have fins that look even more like birds' wings. Underwater they gather speed, then jump clear into the air, unfurling their huge pectoral fins and holding them still, without flapping, while they glide for tens or even hundreds of metres. In 2010, Hyungmin Park and Haecheon Choi from Seoul National University in Korea put dead, stuffed flying fish in wind tunnels and found they glide as efficiently as hawks. The reason fish learned to fly was probably to escape from predators. From below, the sea surface acts like a mirror, reflecting light back down, so a waterborne predator won't see a flying fish unless it's a still, sunny day and it casts a shadow. And they've been escaping hunters' jaws in this way for a long time. Flying fish fossils have been found in the same 235-million-year-old rocks as giant ichthyosaurs, which were perhaps the hunters they were trying to avoid.

Then there are fish whose body shape reveals that they try not to swim at all, if they can possibly help it. Deep-sea anglerfish save their energy (there's not much to eat down there) and drift about, only waggling their tails when danger or food is near. Frogfish sit on the seabed doing their best to blend into their surroundings, and if they need to go somewhere they'll stroll ponderously along using their pectoral fins as legs; if it's really urgent they'll even break out into a gentle gallop. And then there are handfish, fingering their way slowly across the seabed around Australia, their pectoral and pelvic fins splayed out and looking for all the world like little hands and feet.[*]

When fish swim in groups, things begin to get complicated. Half of all fish species spend some of their time swimming together. One in four live permanently with other fish throughout their adult lives. Take a herring or a sardine or an anchovy away from its swimming mates and it immediately becomes agitated.

Fish congregate in two main formations. First there are shoals. These are loose social gatherings in which fish mill about together, without paying too much attention to each other. Then there are schools. Shoaling fish can transform

---

* Anglerfish, frogfish and handfish are all members of the same order, the Lophiiformes.

into elegant, spiralling schools when, for some reason, all the fish suddenly decide to swim and turn in tight synchrony. In a school, everyone swims in the same direction, their bodies in parallel. A school might lose its orderly structure, though, and once again form an untidy shoal. For decades, scientists have studied shoaling and schooling fish to try to understand how and why fish behave like this, as they go from one fish, to two fish, to many.

In his later years Konrad Lorenz, who came up with the idea of eye-catching poster colours, devoted himself to studying the social lives of fish and how they hook up to form schools. Instead of fighting each other and arguing over territories, there comes a point when some fish begin to get along, and Lorenz wanted to watch this switch taking place.

In 1973, Lorenz won a Nobel Prize for his work on animal instincts.* He spent his prize money building an enormous aquarium at his home just outside Vienna. It was four by four by two metres (1.2 x 1.2 x 0.6ft) and contained 32,000 litres (7,000 gallons) of seawater, enough to fill more than 300 bathtubs. He stocked it with various coral-reef fish, including dozens of young Moorish Idols, with distinctive white, black and yellow bands. Then, for several years, he spent most afternoons watching them.

After Lorenz's death in 1989, an incomplete manuscript was found in a drawer in his study. It describes in immense detail what he saw during his long vigils watching fish, totalling more than a thousand hours. He gave names to all his fish, and watched as they performed a complex repertoire of gestures. The Moorish Idols would beat their tails at each other or lock jaws and wrestle; pairs

* Lorenz was renowned for his studies on animals that bond with the first thing they see when they're born, a phenomenon known as imprinting.

would race around the tank side by side, or dart at each other before slowly retreating. Lorenz's notes are filled with sketches he drew of the territories each fish carved out around the aquarium. In March 1977, he wrote in his notes, 'Glub and Fris fused their individual territories, but still exclude Bajo, which selectively either attacks Glub or Fris, when either of the two appears in the "wrong" place ... Kuna is still confined to its shelter at the left lateral wall.'

Eventually Glub, Fris, Bajo, Kuna and the other Moorish Idols all worked out their differences and came together to form a single, permanent school, parading around the tank together. On coral reefs the fish perform similar switches from territoriality to schooling, but no one had ever witnessed it unfolding like this. Lorenz acknowledged that even his large aquarium was cramped, but he was convinced his microcosm provided important clues about what fish get up to in the wild, when no one's watching.

Further aquarium studies, often on small, compliant species like Mosquitofish and Zebrafish, are helping to decipher the choreography of shoals. By watching shoals and schools take shape and tracking the movements of individual fish, researchers are discovering how they adjust their positions relative to each other, not getting too far apart or too close; when predators attack, fish swim in tighter, more synchronous schools with various ways of evading capture: diving to the side, splitting apart and rejoining together in cascades.

Despite appearances, fish do not join to form egalitarian super-organisms, entities with a mind of their own and no leaders guiding the way. Studies are finding there are in fact bold pathfinders prepared to take on the greater risk of getting caught at the front of the school. Hungrier fish also tend to be the ones up front, where they have a better chance of finding food compared to followers, trailing further behind.

Shoaling studies show the benefits fish can reap from swimming together. Most obviously, they avoid predators by confusing them with a blur of identical bodies, and diluting the impact a single hunter can have on a large group. Fish will even take turns to briefly leave the safety of the shoal, swim up to a nearby predator and check out what they're up to. They then return to the shoal and apparently inform their shoal mates if an attack seems imminent, so they can swim away, or if the predator is otherwise engaged. Shoaling fish also boost their chances of finding food, especially if it's patchy and hard to come by; with more fish looking, they'll do better at finding those patches.

Fish also save energy by swimming in shoals rather than on their own. Like cycling pelotons and cars driving in the slipstream of the vehicle ahead, fish further back in a school use less power to keep up. As they swish their tails, fish fling spinning vortices into the water behind them, which their schoolmates have to swim through. But instead of struggling against a turbulent wake, a fish puts itself behind and in between two other fish in just the right spot to get an extra push from these vortices. Even fish at the front of the school make energy savings by riding the bow wave pushed ahead by fish behind them. And as is so often the case, people are learning from nature and applying these fish movements to the human world. Placing wind turbines in similar positions as schooling fish can make them up to ten times more efficient.

To probe deeper into fish-school dynamics, researchers make their own shoals. Based on observations of living fish, they programme fish avatars with rules on how to move. They then release these computerised fish into virtual aquaria and watch them swim around. Virtual predators are released, too, to chase after them, with their own set of instructions on what to do.

Hundreds of computer models like these have been made and refined until, as far as statistical tests are concerned, they're identical to the way real fish move. But is that enough to say a model is true to life?

It's a question posed by James Herbert-Read and colleagues at Uppsala University in Sweden. They set out to see if people could tell the difference between real and computer-generated shoals. In 2015 they built a simple online game with pairs of videos showing green dots swirling around. One video is the two-dimensional trace of a real fish shoal, the other is the output from a model. Players were asked to pick out which they thought was the genuine shoal.

Academics who study fish motion were, perhaps not surprisingly, very good at the game (not to mention immensely competitive), mostly choosing the right shoal each time. Almost 2,000 members of the public also had a go. Although not quite as good as the experts, the gamers could tell that something was wrong and spotted a clear difference between the two shoals, but they weren't always sure which was the simulation and which the real thing.

The computer model failed to pass this fishy version of the Turing Test.* Herbert-Read's games weren't testing the intelligence of simulated fish, but their abilities to swim like real ones. Seeing shoals through many pairs of human eyes revealed that something wasn't quite right with the models, even though statistically they were a good fit. The perfect computerised shoal is still some way off, and fish are still holding onto some of their secrets of how they swim and school and shoal.

---

* In 1950, mathematician Alan Turing developed a way of testing artificial intelligence based on whether a person can tell if another human or a computer is answering their questions, via a screen.

Shoal searching

Many years ago, I set out to uncover a part of the picture of how fish move and shoal. It was an adventure that began one night as a small boat left the northern coast of Borneo. I was a PhD student, and part of a small research team heading to a remote island in the South China Sea. That first night I was too excited to sleep. I stayed on deck as the captain picked a route through the lights of oil rigs, and I watched as the dark shadow of the mainland sank behind the horizon. Then after two days and two nights of motoring slowly through heaving waves, I was too seasick to sleep. Feeling wobbly and ocean-weary, I finally saw the lights of a tiny island peep into view, and an unsettling feeling began to seep in. I'd been planning and anticipating this trip for months, but as I neared my destination I suspected I'd made a terrible mistake.

Swallow Reef is a coral atoll shaped like a teardrop. Above the water line there's little more than a narrow strip of sand and concrete, 1,500m (1,640 yards) long, with a runway servicing a small dive resort and a Malaysian military outpost. The expedition team wouldn't be staying at either but sleeping under the stars or, when it rained, inside a rusting shipping container perched beside the runway. Washing facilities consisted of buckets of water, and to go to the toilet we either crept into one of the sparse shrubs growing at the far end of the island or, preferably, jumped in the sea. There was no internet, no phone signal and little electricity.

Arriving in this place that I'd thought of and talked about for so long, I was excited but at the same time felt a deep shock of disconnection, suddenly realising how far I was from familiar people and places. I was supposed to stay on the island for three months, and I began to doubt if I could see it through.

The following day, things got worse when the team set off to dive for the first time. We left the calm lagoon, passing through a channel cut into the reef and out into the open rolling sea. Inelegantly, I clambered into my scuba gear while sliding from side to side on the tipping deck. It didn't help when the boat's captain announced he planned to keep the engine running. Normally, for safety's sake, dive boats are put in neutral so there's no chance a diver in the water will get lacerated by the whirling propeller.

'It's too dangerous to stop,' he yelled.

The waves were pushing us closer to the sharp, boat-wrecking reef crest. It meant that my dive buddies and I would have to imitate Navy SEALS and execute what's technically known as a 'negative entry'. I couldn't linger on the sea surface to gather my thoughts and check my dive gear was working; just fall backwards off the side of the boat and sink straight down, hopefully beyond the propeller's chop.

I was seconds from screaming back at the captain, giving up on the whole thing and finding a way to get back home. But I dredged up a final burst of resolve, scrambled over the high gunwale and dropped into the water.

Instantly, I passed from hell into heaven.

The water was so clear I barely noticed it was there. It was the closest I've ever come to flying. Below me, the reef spread into the distance. It looked like a garden blooming with flowers, mosses and lichens with scarcely a bare patch of empty seabed visible. This was the healthiest reef I had ever seen. Throngs of fish roamed all around me and among them I spied one of the animals I'd come all this way to see. And, just like that, my fears and worries blinked out.

Humphead Wrasse can be very difficult to find. Most of the time they live on their own, ranging over wide tracts of coral reef. There are a few places in the world, though, where encounters become more predictable.

The humphead I saw on that first dive was a female. I guessed she was about 50cm (20in) from nose to tail; if she'd let me, I could have tucked her neatly under my arm and carried her off. Her flanks were pale green-grey and her tail trimmed in yellow. Her forehead was not especially pronounced or humped; that would come later.

She was swimming intently along the reef towards a particular spot that I would visit the following day, and most days after that for the weeks and months that followed. There I saw not just one wrasse but dozens. Most were females of a similar size, plus one tremendous, presiding male. He was so big he would have had a hard time fitting into a bathtub. He was a similar colour to the females except for his bright blue face and lips, and had a big bump on his head. Occasionally he swam right up and looked me closely in the eye – wondering, I suspect, if I needed to be chased off like the lower-ranking males that milled nearby. This giant male humphead was the first fish I'd ever seen that made me feel I was being carefully and thoughtfully watched.

The humphead shoal formed every day for a week around new moon for a single purpose, an act that for each female would last approximately four seconds. The dominant male, on the other hand, was in it for the long haul. When he wasn't chasing off intruding subordinates, he was busy doing his best to tempt females to join him in the open water above the reef. When she decided the time was right, each female swam upwards with the male trailing eagerly behind. Only then did the size difference between the genders become obvious; he was at least three times bigger than his petite partners. They swam side by side, the male brushing his chin against her body. Then with a swift shimmy she released a cloud of eggs into the water and he added a puff of sperm. The two fish then parted; the female peeled away and swam off back down the reef, and the male turned around to pick up the next member of his harem. One by one, he mated with them all until there were no more females left.

Lots of fish species meet up in shoals to spawn, often in precisely the same places and times each month and year. In the Northeast Atlantic, from the Barents Sea to Iceland and the Faroe Islands, slender cod called Blue Ling gather together to mate hundreds of metres beneath the waves. Orange Roughy congregate on underwater mountains for the same reason. And on coral reefs, many species aggregate to spawn – groupers, snappers, wrasse, surgeonfish – sometimes swimming for days and weeks, across hundreds of miles to reach the spawning site.

Mating shoals range in size. There are small, select groups like Regal Angelfish. A single male, adorned in striking yellow, white and blue stripes, gathers a harem of three or four females. Each evening, 15 minutes before the sun sets over the coral reef, he begins nuzzling the females and leads them one at a time in a spiralling, upwards dance. As they spawn the male flicks his tail, whipping the eggs and sperm into a toroidal vortex – a spiralling doughnut shape, like a smoke ring – which rises up in the water, away from the many mouths on the reef that would gladly feast on the nutritious cloud.

Fish can also reproduce on a truly spectacular scale. Thousands of millions of Atlantic Herring rendezvous on Georges Bank, a shallow sandy shelf between Cape Cod and Cape Sable Island off the northeast coast of the United States. Shoals form at sunset as scattered herring draw together. When a certain density is reached (one fish in roughly five cubic metres (175 cubic feet) of sea) a chain reaction sets off and the shoal grows outwards in waves, similar to the Mexican waves of panic that sweep through a hunted sardine school. The waves move at 60km per hour (37mph), much faster than the herring themselves can swim. A colossal shoal, some 40km (25 miles) across, then begins a slow procession to the southern reaches of the bank, led by smaller groups of fish that seem to know where they're going. On reaching their destination spawning begins, and the water becomes cloudy and thick with the next generation of herring. When the morning comes, it's all over and the shoal disbands.*

Fish gain various benefits from mating this way. Rather than simply hoping they'll bump into a suitable mate somewhere in the wide ocean, it makes sense to arrange a certain place and time to meet. Doing so can also reduce the chance that all those precious eggs will be guzzled by predators. In the Persian Gulf, Mackerel Tuna aggregate and spawn under oil platforms and Whale Sharks show up, too, to sieve the water for their eggs. Even a herd of a hundred Whale Sharks can't eat them all and when they swim off, bellies full, there are still plenty of eggs left to start the next generation of tuna.

Other hunters show up at spawning sites to eat not eggs but the spawning fish themselves. A lot of sharks live on Fakarava Atoll, in the Tuamoto Archipelago in the middle of the Pacific. Divers regularly encounter 600 Grey Reef Sharks in one small area of reef, by far the greatest density of reef sharks documented anywhere in the oceans. Johann Mourier from Macquarie University in Sydney goes with colleagues to study this shark jamboree and they've discovered the ecosystem is flipped upside down. Instead of having more animals towards the base of the food web – the usual way of things – the reef at Fakarava is top heavy with apex predators. Normally, sharks endlessly roam large areas to find enough food, but this unrivalled crowd stay put, for a while at least, because plenty of food comes to them in the form of large, mottled fish called Camouflage Groupers. In June and July each year tens of thousands of them congregate at Fakarava to spawn, and a

* We know all this thanks to a new sonar setup called Ocean Acoustic Waveguide Remote Sensing, or OAWRS, which surveys a patch of sea 100km (62 miles) across every 75 seconds in three dimensions.

lot of them end up getting eaten by sharks, although not so many that the grouper population dwindles. There was probably a time when there were many more great shoals of spawning fish, in other parts of the ocean, being hunted by packs of sharks. But away from remote atolls like Fakarava other hunters have got to them first.

Human fishers have learned to target spawning sites. It's logical to visit the spots where fish congregate like clockwork, usually when the moon is full or new. But unlike sharks, people often take things too far and wipe out whole shoals. In the Caribbean, tens of thousands of Nassau Groupers used to spawn in huge aggregations, but they've been so heavily overfished the majority of the spawning sites no longer form. Similar stories play out for numerous species worldwide. The lost shoals never seem to return, perhaps because young fish learn from the older ones where to go. When those wise old fish have gone, memories of the spawning site are lost with them.*

It was with this in mind that I went to the South China Sea and Swallow Reef to find many Humphead Wrasse. I wanted to work out just how vulnerable they would be if fishers did target their spawning sites instead of catching them one by one across a reef.

Throughout the Pacific, Humphead Wrasse have traditionally been a much cherished fish. In Micronesia and the Cook Islands, they used to be reserved for royal feasts. In the Carteret Islands in Papua New Guinea, only elders are allowed to eat them. Spearing Humphead Wrasse was once an important ritual for young boys reaching manhood on the island of Guam. More recently, though, traditional fishing has been replaced by regional commerce. Seafood

---

* There are signs that Nassau Grouper spawning aggregations are very slowly recovering in the Caribbean, following their protection. In the US Virgin Islands, the depleted Nassau Groupers may be following the more abundant Yellowfin Groupers to their spawning sites.

enthusiasts in Asia have adopted Humphead Wrasse, and various groupers, as their species of choice. Fishers throughout the Indian and Pacific Oceans now target them, often diving down and breathing compressed air through hose pipes that snake to the surface. They take with them plastic bottles filled with cyanide solution, which they squeeze into holes in the reef to stun the hiding fish, incidentally killing other reef creatures, including corals. The idea is to keep the fish alive and send them to cities where people pay top dollar to eat them. The humpheads are displayed in aquarium tanks in restaurants, where affluent diners point to the fish they want to eat. In China, the males' big blue lips are considered a delicacy. And with so much demand, it's taken only a few decades for the Humphead Wrasse to become highly endangered.

At Swallow Reef I wanted to work out what happened when the humpheads met up to spawn. If individual females spawned just once and new fish formed the spawning shoal each day, then there must be a large group of them living around the reef. But if the same fish were showing up day after day, then it would mean a much smaller collection of hard-working females bear the duty of spawning the next generation. It would also mean there was a greater chance that if fishers were to target the spawning site, the entire adult population could quickly be gone.

My task was to learn to recognise individual humpheads. I couldn't very easily round up these big, endangered fish and fix identifying tags on them; instead I kept my distance and took photographs of their natural markings. Their other common name, Maori Wrasse, is a nod to the labyrinthine motifs on their faces that some say resemble the moko tattoos of indigenous New Zealanders.* The iridescent blue lines on a Maori Wrasse's face weave in

* Another common name for them is Napoleon Wrasse, and I've still not found out why.

and out, hopping and jumping in dots and dashes, perhaps a form of poster colours signalling to other wrasse.

I wanted to know if these markings were unique to each fish. If they were, I could use their patterns to recognise individual fish, to track them at the spawning site, to decipher their mating rituals and see how frequently each female came back.

First, though, I'd have to spend many hours in the water with them, catching on camera their complex faceprints.

Throughout my study following the movements of those big spawning fish, I knew they wouldn't roam too far. Adult Humphead Wrasse always need a reef beneath their fins, and won't spend long in open water. Plenty of other fish are less devoted to one particular place, and routinely set off on epic journeys.

Until recent decades, the main way of finding out where fish go was to catch them, mark them in some way, usually with a numbered tag and a 'Return to sender' message, then let them go and hopefully someone will catch those same fish again, somewhere and some time later. Like sending a message in a bottle there's no guarantee the tagged fish will get picked up, and even when they do this approach provides just two pieces of information: the beginning and end point of an otherwise mysterious route. Now, though, fish draw lines on digital maps, with electronic tags following their every move.

Tracking technologies have come a long way since the early days. A Basking Shark was tracked with a satellite tag for the first time in 1982. The shark towed a sizeable floating package behind it on a 10m (33ft) cable; the device relayed its position via satellite whenever the shark swam to the surface. For 17 days, scientists watched the shark's progress from afar as it swam south through the Sound of Bute off the west coast of Scotland, past the Isle of Arran, along the Firth of Clyde and circled around the rocky islet of Ailsa Craig. Then the transmitter broke free, sooner than it was supposed to, and a local resident found it on an Ayrshire beach and sent it back to scientists at the University of Aberdeen. It was still in good working order.

Since then, many large sharks have been tagged with devices now the size of a smartphone, fixed directly to the dorsal fin. A 2017 study revealed the paths of 70 Basking Sharks tracked from Scotland and the North Atlantic on long wintertime migrations. Some hung around the UK and Faroe Islands; some swam into the Bay of Biscay and others swam for months to the coast of North Africa; most covered at least 3,600km (2,200miles).

Research teams have gazed at their computer screens as similar tags have revealed other immense fish migrations. Salmon Sharks from Alaska escape chilly waters to spend their winters in Hawaii. In 2003, a female Great White Shark was tracked 11,000km (6,900 miles) across the Indian Ocean, from South Africa to Western Australia. A photograph of the shark's ragged dorsal fin showed it swam all the way back to South Africa six months later. Bluefin Tuna race east and west along an underwater highway across the North Pacific, between Japan where they breed and California where they feed and fatten up. One young tuna did this three times in 20 months, swimming 40,000km (25,000 miles), equal to the circumference of the Earth. In 2012, the US press stirred panic over health concerns of eating tuna that migrate to American waters from Japan and could have been contaminated by the devastated Fukushima nuclear power plant, even though radiation levels in these fish were so low that eating a normal banana was more dangerous than a tuna-steak dinner.

Besides knowing where a large fish is at any given time, electronic tags are uncovering a wealth of other details of journeying fish. When Great Whites swim across whole ocean basins they commonly pass through wide expanses where there's little for them to eat. Tags that track not only horizontal movements but also depth suggest that, as migration proceeds, Great Whites begin to sink, probably because they're using up the buoyant, fatty reserves in their livers that can make up a third of their body weight. A Great White Shark liver weighing almost half a tonne contains 400 litres (90 gallons) of oil, storing 2 million kcal of energy (roughly equivalent to 9,000 Mars Bars). Like a camel's hump, Great Whites seem to use their livers as a food source to survive long treks through ocean deserts.

Satellite tags have also helped solve a mystery surrounding the brains of manta rays. In 1996, researchers unexpectedly found what seemed to be a brain-warming device in Giant Oceanic Mantas as well as in their relatives, the Chilean Devil Rays. Various sharks, marlin, sailfish and tuna have similar bundles of blood vessels – known as *retia mirabilia,* Latin for 'wonderful nets' – which transfer heat generated by powerful swimming muscles to the brain and eyes, raising their temperature by 10–15°C (18–27°F) above their surroundings and keeping them sharp as they go on hunting raids into deep, cold waters.

Most fish have cold bodies, because seawater saps their body heat as it courses through their gills. Except, that is, in strange-looking fish called Opah. These giant silvery discs have white spots, red fins, a golden ring around each eye and *retia mirabilia* in their gills. Cold blood flowing from the Opah's gills is warmed by blood flowing back from the heart (a so-called counter-current exchange), making them the only known fully warm-blooded fish; these deep-diving predators are the only fish with warm hearts.

It had been assumed that mantas and devil rays were tropical, shallow-water species with little need for warming their brains. One idea was that their *retia* were actually keeping their brains cool. The puzzle was at least partially solved in a 2014 study that began a long way off the coast of Portugal. Thirteen devil rays were tagged at Princess Alice Bank, an underwater mountain in the Azores. The tracked rays swam thousands of kilometres south and plunged thousands of metres beneath the waves, on deep forays that no one had previously known about. The rays reached a depth of almost two kilometres (1.2 miles), placing them among the deepest diving ocean animals.* Again and again, the devils swam straight down, then for an hour or so slowly returned to the surface, probably feeding as they went on layers of plankton. Occasionally the devil rays stayed deep for 11 hours at a time. Why they do this is not exactly clear, but at least now their brain-warmers make sense, as they regularly spend time in waters colder than 4°C (39°F).

Although tagging has provided rich insights into the lives of fish, there are some who call for caution on the indiscriminate use of these technologies, and question who should have access to the information generated, including near real-time data on where fish are. In the American state of Minnesota, a group of anglers recently petitioned to be allowed to use radio-tracking data on Northern Pike, one of their favourite targets.† Scientists collect these data using public funds and so, the fishers argued, the public should be allowed to use the data however they chose. That case proved unsuccessful but another, in Australia, saw shark-tagging data used for a very different purpose to the one originally intended. Following a series of fatal shark attacks

* The deep-diving world record is currently held by a Cuvier's Beaked Whale, which reached 2,992m (9,816ft or 1.9 miles).

† Another electronic tracking technique involves smaller, cheaper devices, compared to satellite tags, which send out radio signals that are detected by receivers fixed in areas of interest, e.g. along a river.

on people swimming off the Western Australia coast in 2014, the government introduced a kill order. Satellite tags had been deployed in a study of shark ecology aiming to help protect these animals from going locally and globally extinct. However, as part of the permit system, all data gathered by tags have to be made available to the licensing agency, and those same data were used to locate and kill sharks.

But on the whole, despite rare instances of being co-opted for nefarious purposes, tagging studies are uncovering great feats of fish migration. Legions of electronic devices are showing that the world's waterways and vast open oceans are thoroughly busy, criss-crossed with well-worn paths between popular places, where animals return year after year, to mate, to feed and to track the passing seasons.

To find their way around and undertake these journeys without getting lost, fish have at their disposal a toolkit of finely tuned senses. They have good vision, they can smell and hear and feel currents passing over their body, and some seem to have an extra sense that remains somewhat enigmatic.

It's not known exactly how, but lots of animals including fish navigate the world with the aid of an inbuilt magnetic compass.* Stretching across the Earth is a net of magnetic field lines that rise in the southern hemisphere and fall in the north and vary predictably from place to place. By detecting the intensity and inclination of local magnetic contours, it's possible to work out where you are in the world. Newly hatched European Eels follow their magnetic sense as they swim from the Sargasso Sea, in the far west Atlantic, towards the Gulf Stream as it flows eastwards and

* Other animals with a magneto-sense include sea turtles, nematode worms, spiny lobsters, homing pigeons, ants and honey bees.

pushes them all the way to Europe. American Eels begin the same journey, riding the Gulf Stream, but they leave sooner and swim west. Migrating salmon seem to remember the specific magnetic field they encounter when they leave their home river and taste salt water for the first time. Several years later, after growing up at sea, the salmon follow this mental magnetic map and return to that same coastal area, then press on inland and eventually smell their way back to spawn in their natal stream.

Small-scale magnetic anomalies can also provide local waypoints. In pioneering tracking studies in the 1980s, American ichthyologist Peter Klimley tracked Scalloped Hammerhead Sharks swimming repeatedly between an island and an underwater seamount off the coast of Baja California. The sharks swam at night through pitch-dark seas and followed dead straight lines. Klimey deduced the sharks were following magnetic gradients that flow around the seamounts, which are made of volcanic basalt and are mildly magnetised as a result.

It remains to be discovered how exactly sharks, salmon, eels and many other animals pick up magnetic fields. It's possible that sharks use their electric senses. When seawater flows past the Earth's magnetic fields it induces weak electric currents, which sharks may detect with electro-sensitive pits on their snouts called ampullae of Lorenzini.* For other species a long-standing theory suggests that some sort of sensory cells, perhaps rich in iron, can detect magnetic fields and trigger nerve signals to the brain. A clue came in 2012, when researchers discovered what appeared to be magneto-receptor cells inside the noses of Rainbow Trout. In a rotating magnetic field these cells swivelled about, like a school of fish all turning together in the same direction.

---

* Named after the Italian scientist Stefano Lorenzini, who first found them in 1678.

Whatever tools they use, there's no doubt fish are master navigators that can find their way not only across entire oceans but also whole continents. A swathe of South America, where the Amazon River and its tributaries percolate through dense rainforest, is home to thousands of freshwater fish species, including a giant catfish known in Portuguese as the Dourada. These enormous fish can grow to almost two metres (6.5ft) long; they have a wide mouth, long whiskers and smooth, shiny skin that makes them look as if they've been dipped in mercury. Together with several close relatives, these catfish support the biggest fisheries across the Amazon Basin, and fishers have long known there's something special about them – these fish are great wanderers.

Compared to other more globally famous fish species, especially sharks, Amazonian catfish are rather overlooked, and scientists lack resources to study them. The catfish have never been fitted with electronic tags; instead, a few teams of dedicated researchers have turned to simpler but more labour-intensive methods to find out where catfish go. One group gathered decades of data from interviews with fishers and surveys of catfish, young and old, across the river basin. Another team bought fish from markets around Amazonia, in the cities of Manaus and Belém, and from inside their heads picked out small ear stones, otoliths, which help the fish balance and hear. As the fish grew, their otoliths became imprinted with chemicals from the waters they swam through. And because the water's chemical composition differs from place to place, it's possible to decode the layers of chemical traces and reveal where a fish has lived at different stages of its life.

These studies are piecing together the catfish's story and confirming what fishers suspected. Dourada and other giant catfish venture on spectacular migrations. Their lives begin far to the west in the headwaters, high in the Andean mountains. The young larvae then drift with the currents eastwards, arriving perhaps a month later at the mouth of the Amazon River, close to the Atlantic on the opposite side of the continent. There they wait, hunting and growing, for at least three years. When the rains come and the river floods, the full-grown catfish gather in great shoals and set off for the west, swimming upstream through white waters and back to the mountains. Genetic studies also hint that, like salmon, Dourada could make their way to spawn in the very same rivers where they were born.

The return trip across Amazonia, between the mountains and estuary, is close to 12,000km (7,500 miles), as far as New York to London and back again, or John O'Groats to Cape Town. This is the longest swim of any animal through freshwaters alone. Why the catfish go to all that effort and swim so far is another fish mystery that, for now, remains unsolved.

Big fish, past and present

Diving every day with Humphead Wrasse at Swallow Reef in the South China Sea, I did my best to pretend I wasn't there, which was difficult when there was nothing to hide behind. At first the spawning fish had been wary of me and rushed away before I could get good pictures of their faces. I feigned disinterest, turning my camera away so the domed glass lens wouldn't convince them they were being eyed by a giant predator. Gradually they grew less camera-shy, and I worked out where to position myself. Hovering at the edge of the male's territory, females would pass me after they spawned as they headed back home. Best of all was when they swam head on, then turned at the last minute to avoid bumping into me, providing a perfect shot of one cheek, left or right.

Between dives I had a lot of spare time on my hands. I wrote long letters home and persuaded pilots of the dive resort's light aircraft to post them for me back on the mainland. I thought a lot about my best friend who was off on his own adventures, in the dry forests of Madagascar, and had no clue that less than a year later we'd be planning our wedding. Out on the island I didn't look back at the digital pictures I was shooting underwater. The electricity supply was patchy and I needed all the power I could get for recharging camera batteries rather than running my laptop. So it wasn't until I was far from the ocean, back in the lab in England, that I began to gaze at hundreds of fish faces.

I sifted through the images, tracing shapes, often getting lost along the meandering facial patterns as I tried to decipher the details of the spawning shoal. Eventually I picked out two pictures, shot on two different days, and saw the same face. Both showed the same three dark lines running back from her eye, the same white speckles on her forehead and a maze of golden cheek-scribbles. This female wrasse had come back to spawn two days in a row. My fishy game of snap was going to work.

From my growing catalogue of fish portraits I discovered the same females regularly spawned day after day and I estimated there were roughly a hundred mature females on the reef. Given a chance, those fish would have continued to spawn at Swallow Reef for many years, flinging ever more eggs into the oceans. And in time they would have returned to that same site in a very different role.

Humphead Wrasse are one of many marine fish that undergo a spontaneous sex change. Many are born as females, then later in life, when they're at least five years old, become males. Their ovaries shut down, sperm-producing testes stir into action and they grow a big bump on their heads. From then on, the fish's task is not to add eggs to the mix but to find their place in the hierarchy of males, and do their best to one day take over the spawning territory. Until then, they'll sneak in and mate with females when the main male's back is turned.

A similar gender switch happens in other wrasse, in cichlids, parrotfish, groupers, gobies and bass. Some fish do things the other way round, starting as males and becoming females, including the Orange Clownfish, the famous fish otherwise known as Nemo. Had the Disney movie been biologically correct, when Nemo's mother went missing, his father should have changed sex and taken on the role as dominant female among the anemone's stinging tentacles (and Nemo himself wouldn't have been living with his dad in the first place; after they're born, young clownfish don't stay at home but drift off to find another anemone).*

Some fish switch sex in both directions. Chalk Bass in the Caribbean have both male and female sex organs at the same time, and use both throughout life-long partnerships. Pairs of bass live together, often inside old conch shells, and swap gender roles, male to female then back again, up to 20 times a day.

But none of my female Humphead Wrasse transformed into males. After I left Swallow Reef things changed. I had hoped to go back but the research team didn't re-form. Only last year, I saw this part of the South China Sea again, although this time from the air. On a flight from Singapore to Manila, the pilot pointed out that we were flying over one of the islands China has claimed as its own. 'Today the air is clear and you can see the runway,' he said. I looked down at an island very like the one I'd lived on for months, only with a flotilla of large, forbidding ships moored around it.

In the last few years, China has been aggressively expanding its presence into these long-disputed waters, pouring sand and concrete onto coral reefs to build artificial islands and military installations, reinforcing sweeping claims over most of the South China Sea. That's despite competing claims among several other sovereign states

* But maybe I'm just spoiling things.

whose shores are lapped by these waters, including Malaysia, Vietnam, Taiwan and the Philippines. The US is also deeply involved in this distant sea which is rapidly becoming a global security hotspot. It's no great surprise, then, that amid the mounting geopolitical tension no one was going to miss a few fish.

I stopped in the Philippines for just long enough to catch a connecting flight and continued my journey eastwards another 400km (600 miles) to Palau, a cluster of forested islands out on their own in the western Pacific. There I met Lori and Pat Colin, founders of the Coral Reef Research Foundation, who've spent years studying Palau's reefs, including a substantial population of Humphead Wrasse. Over dinner, they confirmed a rumour I'd heard. Lori and Pat had been to Swallow Reef a year after I finished my PhD, and they didn't see a single humphead.

The atoll was never officially a marine reserve, although the Malaysian military had acted as *de facto* protectors of the reef, ensuring no other vessels came close. It was out of bounds except to a few divers and occasional researchers. Lori told me the story she'd heard, that fishermen had been allowed onto the reef. It was one of the fleets that roam Southeast Asia, hunting for the last of these valuable fish, and in Swallow Reef they had taken all the Humphead Wrasse they could find. Had the fishers gathered them from the spawning site? Perhaps. The humphead's former abundance on the atoll had been common knowledge, long before our expedition and my PhD. Still, I couldn't help wondering if our attention and presence had fuelled interest in those fish.

My efforts to study humpheads suddenly felt hollow. While I'd been focusing in on the minutiae of their lives, bigger forces were at work. All those fish I'd learned to recognise from their facial patterns had been doomed from the start because of the price tag on their heads. I had witnessed and documented a phenomenon that may never happen again, not there anyway.

Swallow Reef wasn't the last place I saw Humphead Wrasse. Diving in Palau I saw them almost every time I went in the water. I saw huge males cruising around and usually two or three large females on most reefs. I saw hand-sized teenagers and even thumb-sized juveniles flitting through shallow lagoons. And at the end of one dive, as I was waiting out my safety stop at five metres (16ft), I looked across and saw a pair of humpheads just below the surface moving in a familiar way. They shivered their bodies together and released into the water a milky cloud, exactly what these big fish should be doing.

As well as humpheads, I spotted several hundred other fish species in Palau (out of roughly 1,400 known to live there). There were big, old fish with wrinkles on their skin who've roamed these reefs for decades, and on every dive I saw sharks. It was inspiring to witness this spot in the ocean that's being so carefully looked after. Almost half the waters close to shore are safeguarded inside a network of reserves, based on a local, centuries-old custom of halting fishing in certain areas to let stocks replenish. Offshore, away from the islands, 80 per cent of Palau's maritime territory was set aside in 2015 as an ocean sanctuary. Industrial fishing fleets, mainly hunting tuna, are no longer allowed and Palauan fishers are the only ones who now dip into the remaining 20 per cent. The result is that a riot of fish is still here, living long lives in huge, healthy populations, and a lot them assemble in colossal shoals to mate. Studies in Palau are revealing much that wasn't previously known about the ways these fish come together, when there's enough of them to form spectacular spawning shoals.

In the office of the Coral Reef Research Foundation, Pat Colin shares with me stories of his life spent watching fish and deciphering their intricate mating rituals. With his snow-white beard and a glint in his eye he's clearly, after all these years, still devoted to the underwater world. He browses through hundreds of computer files and picks out videos to show me of shoaling, mating fish.

'Isn't that lovely?' he says' as we watch pairs of parrotfish swim cheek to cheek in languid circles, spiralling slowly down through the water. 'It's a dance. I say it's tender. It's the only word I can think of to describe it.'

Next Pat shows me a video shot at Blue Corner, one of Palau's most famous and impressive dive sites. In the water are hundreds of Moorish Idols swimming synchronously just above the reef, all turning together with flicks of their trailing, ribbon fins. It makes me wonder what Konrad Lorenz would have made of them, following his studies of just a dozen of these fish in his Viennese aquarium tank. 'There are a lot of good naturalists who send me videos and pictures,' Pat tells me. More divers than ever are in the water in Palau. Dedicated dive tours have begun taking visitors to the right spots at the right times to watch the spawning pageants play out before them. 'Things like this get noticed,' Pat says, even though Moorish Idols only form shoals for a few days every year.

'Do you want to see something really cool?' Pat asks me. Of course I do. He clicks another video file, this one labelled *Lutjanus fulvus*, the Blacktail Snapper. 'The fish are really incredibly shy,' he says. 'You can't get close to them.' And he doesn't try to. Instead, he fixes GoPro cameras to the reef and programmes them to take one shot every minute for a week at a time. These tiny, tough, waterproof cameras were developed for extreme sports fans; the internet is full of videos shot by surfers, skydivers and skiers with cameras strapped to themselves. But they've also been adopted as scientific spy cameras. In this way, Pat has eavesdropped on the mating shoals of Humphead Wrasse and another local giant, the Bumphead Parrotfish, which come together in their hundreds to spawn.

On the computer screen I see a panoramic view across a coral-rich reef from Pat's three cameras. He hits play and we watch as one or two fish are caught on camera, swimming by. Then, quite suddenly, first one then all three frames are filled with yellow stripes and dark tails.

Snappers block the reef and each other from view until it's impossible to count them. Now and then a big eye gazes close into the lens, then disappears again in the next shot. The time-lapse images scroll through and for a few moments, the equivalent of an hour in real time, fish cover the screen. 'Then they're gone,' says Pat. And as fast as they appeared, the spawning snappers disperse and the reef goes back to the way it was.

Osiris and the elephantfish

Ancient Egypt, 2,400 years BP

Stories carved inside Ancient Egyptian pyramids tell of a great and wonderful king called Osiris who ruled over Egypt with his wife Isis. Everyone adored them both except for Seth, Osiris's evil, jealous brother, who plotted to bring him down. One year, at the feast for Osiris's birthday, Seth brought in a trunk inlaid with gold and jewels. 'Whoever fits inside this chest,' Seth announced, 'will show themselves to be the most true and loyal person.' The courtiers and partygoers all took their turn to climb into the chest, but they were either too big or too small. Osiris himself stepped forwards and showed that he was a perfect fit. As soon as Osiris was inside the trunk, Seth's henchmen slammed the lid shut and nailed it down. Then they took the trunk and threw it in the River Nile, and Osiris drowned.

Lamenting her lost husband, Isis went to the river and searched for his body, finding that it was already breaking apart. With her magic she restored Osiris and later bore him a son, Horus, who became the god of the sky.

When Seth heard about what Isis had done, he feared Osiris would return to seek vengeance. So he went to the river and found the remains of his brother's body, then cut them into fourteen pieces and scattered them across Egypt, to make sure Osiris would never return. Once again enraged by Seth, Isis travelled across the land to gather up all the pieces of her husband, but she only found thirteen. The fourteenth piece was his penis. She couldn't find it because an elephantfish, known as Oxyrhynchus or Medjed, had eaten it.

From then on people worshipped the fish that had eaten such an important part of their great king. They believed the fish was a divine manifestation of Osiris, who also became the god of the afterlife, of death and life and resurrection. In temples dedicated to Osiris people left offerings of bronze fish figurines and mummified fish.*

* Tilapia, another fish from the Nile, also came to be associated with Osiris. People saw tilapia spitting out hundreds of tiny fish from their mouths and believed they possessed great regenerative powers; so they became a symbol of rebirth and renewal. Women and children would wear fish pendants as protective amulets. And tilapia, people said, led the boat of the sun god, Ra, on his journey through the Egyptian underworld and back into the sky, where each morning he was reborn.

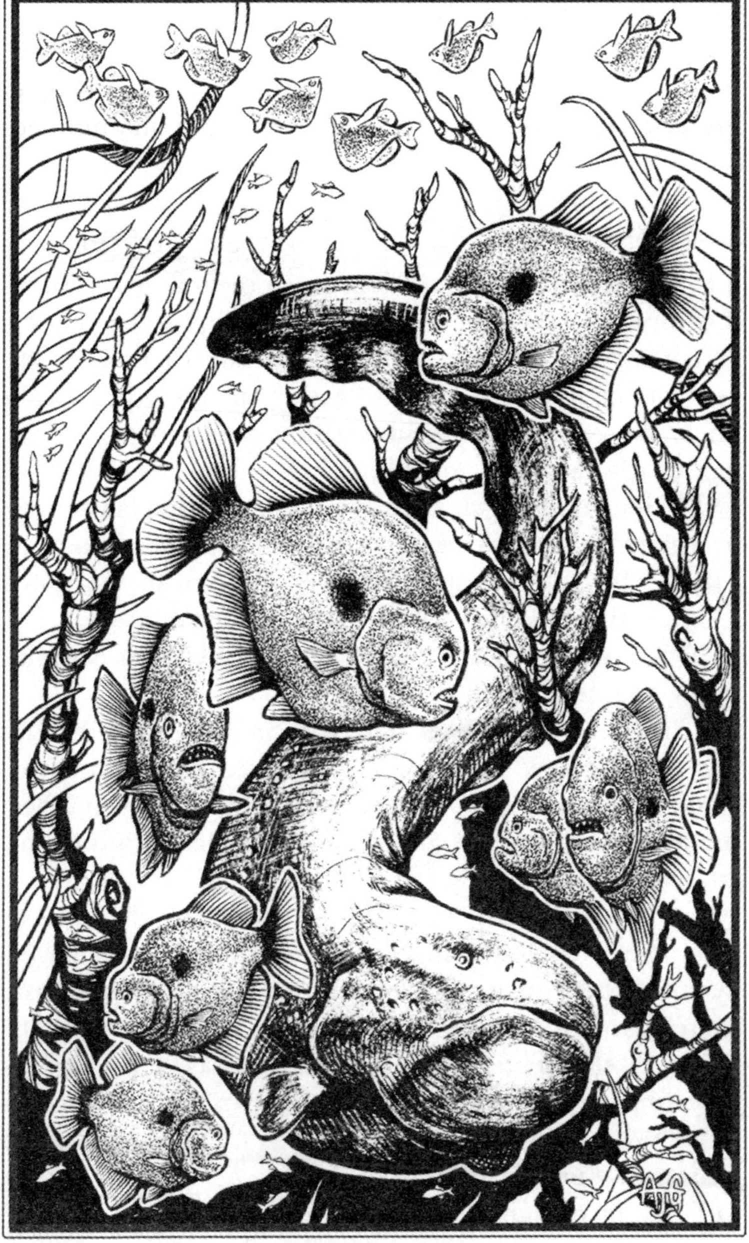

CHAPTER SIX:

FISH FOOD

CHAPTER SIX

Fish food

It's not long after sunrise when I hear the sounds of drums coming from the other end of the beach. I've been lying awake for ages, listening to waves gently tugging at the sand and hoping for a signal to get up and go swimming. I unzip my tent, grab my snorkelling gear and join a few other early risers in a small boat.

The journey is short, just a few minutes motoring around to a narrow channel between this and the neighbouring island. My guide, Semmi, has already been in the water this morning, and he's optimistic for a fruitful swim.

'When we get in, stay close to me,' he calls out over the idling engines.

I drop over the side of the boat. Straight away the fast current pushes me along and I fix my gaze into the distance. There's nothing but blue-green haze. Then I hear a muffled shout, look up and see Semmi with his hand high in the air, palm open – the sign of a manta.

A dark shape glides towards me, just below the surface. The manta ray is swimming upstream but seems unfazed by the current; it is, after all, mostly fin – two pectorals stretched out into triangular wings.

The cluster of humans in the water pauses for a moment, watching, trying to take in its great size. This one is modest for a manta, with perhaps only a two-metre (6.5ft) wingspan, but still, it's improbably big for a fish. It begins to fade out of sight and a silent whistle goes off, stirring everyone into action. The snorkellers frantically paddle to keep up, trailing after the undulating fish, like kids playing soccer, all trying to get the ball at once. Another, slightly larger manta soon arrives. This one has a neat shark bite taken out of one fin, but the injury doesn't stop it swimming smoothly and fast, followed by its splashing, kicking human entourage. Prosthetic, plastic fins are no match for the manta's speed and it's not long before people begin to fall behind and scramble into the boat to hitch a ride back upstream. Again and again the group takes to the current, like a fairground ride of which nobody grows tired.

Each year, around sixty mantas make repeat visits to this channel, midway along an island chain in northwest Fiji. In a similar way to my Humphead Wrasse, individuals can be identified from black and white markings on their bellies. On average, between April and October, three mantas show up every day. On one occasion, there were 14.

These enormous flattened elasmobranchs come here primarily to feed. Pouring through the channel is a seawater soup laden with zooplankton, animals too small to see with the naked eye but so abundant they make the water thick and cloudy. Mantas swim through their food, their huge mouths wide open. Plankton get lodged on feathery filaments called gill plates or gill rakers that fringe each of their ten gill slits. Every so often the manta closes its mouth, coughs and swallows down the zooplanktonic paste.

Mantas aren't the only fish that routinely show up in this Fijian channel to eat. While I paddle around, a seething shoal of mackerel hurries past. Silver bodies streak through the water, forming a restless, shape-shifting mass that sweeps up, dives down and performs tight U-turns, swimming back on itself. Most of the time the bodies within the school are lined up, their dark stripes in parallel. But now and then the group fragments into glinting disorder before falling back into synchronised formation. Trailing the mackerel is a hunter's shadow, a Giant Trevally, attracted to the fizzing activity and waiting for its chance to attack. The mackerel swim on, sifting the water for plankton with their mouths open like a thousand miniature parachutes.

There's another source of food that comes to the channel with the mantas. Finger-sized fish flit over their bodies, picking off parasites and scraps of dead skin. These are cleaner wrasse; every day mantas spend hours getting themselves groomed and cleaned.

While so many fish gorge themselves, the human swimmers in the channel begin to tire and think of their own breakfast. While the boat zips around picking people up, I hang back and spend a while longer riding the current and watching the water around me.

Once more the two mantas emerge into view. They found a good spot, packed with plankton, and stopped to make the most of it. The two fish are spinning pirouettes through the water. They arch their bodies and swim backward loops, chasing but never quite catching their long tails, all the while their gills filling up with food.

Fish get food for themselves in a huge variety of ways. It's one of the things that sets them apart from other vertebrates, which in comparison tend to have rather conservative diets. Fish, on the other hand, will eat almost anything, in almost any way, and this has major implications for the rest of the underwater realm.

Many fish are hunters, some legendarily so. Around 30 species of Amazonian piranhas have gained a fearsome reputation as unhinged flesh-eaters, ready to nip your finger off as soon as you trail it in the water. Such stories were stirred by early western explorers, including US President Theodore Roosevelt. In his 1914 book *Through the Brazilian Wilderness*, Roosevelt describes piranhas mutilating swimmers, and going mad at the mere sniff of blood. More credible are reports that the locals had put on a show for their esteemed visitor. A few days beforehand, they netted a shoal of piranhas and deprived them of food. When Roosevelt arrived a dead cow was thrown in, and the ravenous fish quite understandably had a feast. Piranhas aren't nearly as dangerous to humans as their reputation would suggest. A lot of the most gruesome cases probably involved fish scavenging from bodies that were already dead. There are, though, concerns that an increase in dam-building across their range in South America is creating ideal spawning grounds for the fish. Along with droughts that are pushing them into deeper waters, it could be that piranhas are coming into more regular contact with human swimmers and as a result more attacks are being reported.

Predatory fish can be extremely adaptable. In southern France, where the River Tarn flows through the historic city of Albi, giant catfish have learned to catch pigeons. Originally from Eastern Europe, these metre-long (3ft) fish were introduced into the river in 1983 for anglers to catch, and since then they've settled in well to city life. Known as the Danube Catfish, they lurk in the shallows next to gravel islands in the middle of the river, where pigeons come to drink and clean their feathers. When the birds get too close to the water's edge, the catfish leap out and deliberately beach themselves (some whales and dolphins employ similar tactics). Researchers from the University of Toulouse took turns to watch this from a nearby bridge, and they saw that roughly one in three hits ended in a catfish getting a pigeon meal. It could be one reason why these fish are such successful invaders, modifying their diet to whatever food is available in new territories. The catfish now live in ponds and rivers from the UK to China.

The world's waters are filled with many fish chasing after other animals, but there are also plenty of vegetarians. Some piranha species are gentle herbivores that play a crucial role in river ecosystems; with huge, crushing teeth they chew on large seeds, helping to disperse them and promoting germination. Of around a hundred species of parrotfish the majority eat algae, cropping vegetation with teeth fused into parrot-like beaks. They boost coral-reef health by keeping in check seaweeds that can otherwise quickly overwhelm and outcompete the corals. A 2017 study of a 3,000-year fossil record from Panama revealed close ties between the rise and fall of parrotfish and coral reefs. Katie Cramer from Scripps Institution of Oceanography in San Diego excavated parrotfish's beaky teeth and measured historic rates of coral growth. She found that when the fish were doing well so were the corals; but when the fish declined, including in the last 200 years from overfishing, the reefs stopped growing and seaweed took over. Caribbean reefs are among the most degraded on the planet today. Cramer and her co-authors concluded that the only hope for their recovery is for the 'significant and immediate reduction of fishing on parrotfish.'

Damselfish take vegetarianism to another level. They're one of a small clique of animals, besides humans, to shift from being wandering hunter-gatherers to settled farmers. Ants, termites and beetles farm fungi; deep-sea yeti crabs grow bacteria on their hairy arms; and on coral reefs, damselfish carefully tend their seaweed gardens. In territories set up among branches of dead coral, damselfish weed out inedible types of seaweed and cultivate a lush turf containing only the species they prefer to eat, sometimes a monoculture of one species.* Lacking stomach enzymes for digesting tough seaweed, damsels select the softer, tastier varieties.

A lot of work goes into maintaining these gardens, chiefly in defending them from intruders. Damsels are

* Damsels don't sow or transplant their favoured seaweed species, but wait for them to naturally settle; it's thought early human farmers may have done a similar thing, weeding out unwanted varieties from mixed wild vegetation.

some of the easiest fish to watch on reefs because they're unafraid of just about everything else, people included, and they don't swim away and hide – quite the opposite. Whenever I float above a shallow reef I get accosted by aggressive little fish that are convinced I've come to steal their precious seaweed. I've watched a large shoal of surgeonfish – wide-ranging herbivores – come stampeding past, stirring up an almighty ruckus among the resident damsels. All at once dozens of angry gardeners, who had been hidden, rose up in the water column, shouting and nipping at the surgeonfish intruders. Damsels will even pick up sea urchins by their spines and carry them off outside their gardens. This may seem paranoid, but studies have shown that when damsels are experimentally removed from their gardens, it only takes a day or two for other animals to come in and chomp through their prized specimens. Some of the most delectable seaweeds, including a red variety called *Polysiphonia*, have only ever been found inside the damsel's protected turfs, showing how fish and seaweed have come to rely on each other.

For corals, the damsels' gardens are not such good news. The fish will bite and kill larger corals to make space for their gardens. Individual damselfish territories range from the size of an A2 sheet of paper to a table-tennis table (the damselfish range in size from a finger-length to a hand span). In places where fishing has depleted bigger, predatory fish – the kinds that people like to eat – damselfish have exploded in numbers, and spread their gardens across the reefs. As well as clearing away corals, the seaweed turfs alter the types of microbes in the ecosystem and seem to promote diseases among the remaining, living coral.* The loss of bigger, predatory fish shifts the balance towards smaller fish, like damsels, and indirectly corals suffer.

---

* Diseases can be a big problem for corals; epidemics in the 1980s are another cause of the demise of Caribbean reefs, along with the lack of grazing parrotfish.

Beyond hunting and farming, there are yet more ways in which fish get their food. To get a good idea of just how versatile fish are in their eating habits, a good place to look is the Great Lakes of East Africa, among the flocks of cichlids. They've evolved a staggering range of diets, which it's thought have assisted their speciation and allow species to coexist in the same place by dividing up resources between them.

There are plankton-eaters, snail-, sponge- and leaf-eaters, dirt-slurpers, eyeball-peckers and cichlids with luscious lips to suck insects from holes in rocks. Others get food by pretending to be dead. They lie motionless on their sides, and their mottled coloration makes them look as if they've already begun to decompose. Scavenging fish then come to investigate and get a nasty shock when the apparently dead fish leaps up and takes a bite out of them. And there are predatory cichlids that have learned to dash up and headbutt a female cichlid while she's brooding young inside her mouth. This forces her to spit out her babies, and the intruder quickly swallows them down.

No matter what they eat, all fish face the same challenges of living and feeding underwater. Water is much stickier than air, meaning that when aquatic animals lunge forwards to catch something, their own body creates a bow wave that pushes their food away from them. The wave also alerts prey to an approaching predator, giving them more time to escape. Mantas, mackerel and all the other filter-feeding fish partially avoid this by opening up a big mouth and letting the water flow through their bodies, rather than around them.* Filter-feeding is a way of getting food that

* 3D-printed models of whale-shark gills placed in a flume tank have revealed how their gill rakers don't get clogged up: tiny vortices form, whisking solids off the filter surface and keeping it

only really happens underwater, from paddlefish that sift the lakes of North America to the enigmatic Megamouth Sharks that push their giant maws through deep seas.*

The challenges of hunting in sticky water have also led to a uniquely fishy trait. A common feeding strategy for fish (especially teleosts) is to fling their jaws forwards. This creates a smaller bow wave than moving their whole body, and it can also be much faster. It's how pet goldfish peck at dried food flakes. And if you pull down the lower jaw of a dead salmon or trout, its upper jaw will automatically swing upwards and outwards via a series of articulated bones.

Fish have been thrusting their jaws out for at least 100 million years. In that time they've reached further and further forwards. Today, the winner of the jaws-out contest is the Slingjaw Wrasse, a tropical fish that can extrude its snout into a tube that measures 65 per cent of its head length. If I had the same talent, I'd be able to bite a chocolate bar dangling on a string 13cm (5in) in front of my nose, without moving the rest of my head.

A close second is the Goblin Shark. In French it's known as *requin lutin* (leprechaun shark) and in Spanish it is *tiburón duende* (elf shark). When this species was first discovered in the 19th century it was known only from dead, ugly specimens with jutting, snaggletoothed jaws, like badly fitting dentures. Only much later did it become apparent that living Goblin Sharks tuck their jaws away, making them a much sleeker, less terrifying apparition. They only stick out their jaws when they're eating – or trying to. They were recorded doing this for the first time in 2008, when a team of Japanese scientists netted live Goblin Sharks in a

---

clean. Engineers are hoping to imitate this to prevent clogging in industrial machines that filter beer and dairy products.

* Technically speaking orb-web spiders 'filter-feed', though I'd say this is more of a trap, as most of their prey isn't blown in passively.

deep canyon in Tokyo Bay. The sharks were carefully brought up alive, then filmed while they swam around in the shallows. A couple of times in front of the camera the sharks tried to bite something, including a diver's arm (luckily he was wearing a thick wetsuit). The film showed the sharks' lower jaws dropping open into yawning gapes of 116°,* then shooting forwards like a sling-shot. Goblin Shark jaws reach half their head's length and snap shut in under half a second – the furthest, fastest jaws of all the sharks.

Beyond simply coping with the stickiness of water, many fish take advantage of it and have learned how to suck: they shoot their jaws forwards, puff out their cheeks and gulp in a mouthful of water and with it any prey and particles of food that get swept up in the viscous current (try it yourself, slurping a bowl of minestrone soup). In retalliation, prey fish have evolved a speedy escape response; sensing the bow wave of an approaching predator, smaller fish will lurch forwards by a single body-length, just enough to avoid the predator's short-range suck.

Among the champion suckers are seahorses. Like mantas they eat zooplankton, but rather than filtering seawater, seahorses peck one by one at individual tiny shrimp. A seahorse lines up its snout just below a target, then tightens muscles in its head that stretch tendons, like pulling back an elastic band before you flick it at someone. A trigger muscle then releases the stored elastic energy and the seahorse's snout suddenly rotates and sucks in a shrimp.

Newborn seahorses are especially good at this, an unusual skill in one so small. Most fish take time to learn how to suck. They have to get used to coordinating their jaws and muscles, and their suction generally improves as they get older. Seahorses, though, spend several weeks growing inside their father's pouch and they're born fully developed and

* Compare that to an estimated 63.5° for a *Tyrannosaurus rex*.

ready to go.* High-speed videos shot in 2009 show that the young seahorses pivot their heads three times faster than adults, faster even than a mantis shrimp when it swings its ferocious claws to crack open a seashell. The foals rotate their snouts at the equivalent of 80,000° per second (mantis shrimp manage a mere 57,000°/sec).

Another group of fish does things in reverse. Rather than sucking in sticky water, they spit it out. Seven species of archerfish use water as a weapon. The earliest known written record of their marksmanship comes from 1764, in a letter sent from Dutch East India† to a fellow of the Royal Society in London. The letter accompanied a preserved specimen of a 'jaculator', as archerfish were then known, donated by Governor Hummel from the hospital in the colonial capital Batavia.‡ Hummel had heard about the habits of these curious fish but wanted to see for himself if the stories were true. He ordered several archerfish to be caught and placed in a tub of water, with a stick propped over the side and a fly pinned to its end. How thrilled the governor was when, day after day, he saw his fish shooting the fly and never missing their mark.

Over the years, research has gradually revealed the true extent of the archerfish's talents. They compensate for the way light bends as it passes between water and air to infallibly hit targets up to three metres (10ft) away, like a male Splash Tetra keeping his eggs wet. Then archerfish position themselves in just the right spot to grab the fallen insect. Not only that, but they shoot their in-built water pistols with power five times greater than any vertebrate

---

* Seahorses are the only animals we know of in which the males get pregnant and give birth; females deposit eggs inside the male's pouch, where they hatch and grow, while he nourishes them before giving birth; with sharp contractions of his pouch he squirts baby seahorses out into the world.

† Now Indonesia.

‡ Now Jakarta.

muscle can muster. Until recently it was assumed that the secret to this feat was some sort of catapult, perhaps similar to a seahorse's rotating head. But no matter how closely people looked, they found no trace of such a device in archerfish. It seems that archerfish don't actually have bows and arrows.

In 2012, Alberto Vailati and colleagues from the University of Milan in Italy finally solved the puzzle. Rather than relying on muscular force, these little fish manipulate the water itself. An archerfish spits out a jet of water by pushing its tongue along a groove in the roof of its mouth. Vailati's team worked out that the fish push harder towards the end of a water stream, so droplets further back bunch up and collide with others ahead of them, and they merge. So instead of spraying their target with a drizzle of fine droplets, archerfish strike out with a single swift blob, powerful enough to knock a small creature from its perch. When you or I throw a water bomb it may initially sail through the air towards our victim, but it quickly succumbs to the forces of gravity and air resistance, slowing down and plopping to the ground. Archerfish shoot water bullets that actually speed up the closer they get to their target.

Fish that swim through electric dreams

As well as being sticky, water is also very good at conducting electricity – at least a billion times better than dry air. That's why it's a bad idea to change a light bulb with wet hands. But rather than worrying about the health and safety risks of mixing water and electricity, there are fish that do this on purpose.

For thousands of years, people have known there are fish that have a certain spark to them. In Ancient Greece, doctors placed electric rays on women during labour, apparently to help them cope with the pain. Ancient Egyptians caught electric catfish from the River Nile and may have used them to treat people who were having epileptic fits. And at the turn of the 19th century, the Prussian explorer and naturalist Alexander von Humboldt saw horses and mules being attacked and overwhelmed by eels in a muddy pond in Venezuela. These and hundreds of other fish species share an unusual ability: they make and control large amounts of electricity.

Every living creature is electrically powered. Charged ions flow in and out of cells, in particular nerve cells; conducting messages, contracting muscles, making thoughts (the electricity that comes out of sockets and powers electronic devices is made of flowing electrons, another form of charged particle). In most living bodies the electric charges are tiny. Various fish, however, have evolved organs that accumulate and amplify electricity and send it out in deliberate bursts. Hunting with electricity is something only fish do.

Years ago, as a zoology student, I met an electric fish, a species of elephantfish (from the same family as the Ancient Egyptian story of Oxyrhynchus). I peered at the one I'd been given in the laboratory class, and saw that its snout was neither as long nor as dextrous as an elephant's trunk. It bore a closer resemblance to its common name in German, *tapirfische*, named after the South American Tapir. On closer inspection, I saw that my fish's proboscis wasn't in fact its nose at all, but an elongated chin.

My task was to map out the electric field emanating from the elephantfish by measuring the current with an electrode dipped at intervals around the aquarium tank. The wobbly diagram I drew showed the fish was surrounded by a cloak of concentric lines. This was the electric field generated by modified muscle cells at the base of its tail. It gave out a gentle, constant pulse, not strong enough to zap the finger I waggled in the water when no one was watching.

I was recreating an experiment conducted 50 years previously in the same laboratory in Cambridge's Department of Zoology by the man who uncovered the elephantfish's hidden talent. Hans Lissmann had seen these fish at the aquarium in London Zoo and noticed that they swim backwards without bumping into things. With eyes focused ahead, they can't see behind themselves, so he wondered if they used some other sense to find their way around. With similar equipment to the set-up I was using, Lissmann was the first to detect the elephantfish's weak electric field. He worked out that elephantfish use electricity as bats use sound. This isn't echolocation, though, but electrolocation.

Just as Lissmann had, the next step in my experiment was to stick a glass rod – an electrical insulator – into the aquarium near the fish and once again map out its electric field. This time the lines were distorted as the electric pulse flowed around and not through the insulated rod. My fish would have known the glass rod was there and not necessarily because it saw it. Until a few years ago, elephantfish were thought to be blind, but recent studies suggest they can probably make out large, moving objects thanks to little crystal-filled cups in their eyes that intensify dim light. Even so, these fish are far more sensitive to electricity than to light. All along its body, my elephantfish had dimples that sensed its own electric charge. Shifts in that field would have told the fish there was something new nearby.

Around 200 elephantfish species live in rivers in Africa, where they emit electric pulses to probe the muddy waters and detect distortions in their personal electric field as it bounces off objects around them. They use their long, sensitive chins to search for food hidden in riverbeds. To process all the information coming in from their tingling senses, elephantfish have enormous brains that use up to 60 per cent of their oxygen supply. Theirs is a similar ratio of brain to body size as humans', except we use only 20 per cent of our oxygen to power ours.

The gentle elephantfish lie at one end of a spectrum, with the notorious Electric Eel at the other (a fish that goes by the pleasing scientific name of *Electrophorus electricus*). They're not true eels but one of many species of South American knifefish, and they can generate 600-volt jolts that can incapacitate and even kill other animals. Kenneth Catania from Vanderbilt University in Tennessee has been getting to know Electric Eels in ways no one has before. Among many new insights, his studies have shown that they don't fire off their powerful shocks at random, in the hope of hitting something, but they use electricity in much more subtle and smart ways. An eel will often commence its hunt by sending two or three charges into the water. If small fish or crustaceans are hiding, they will give themselves away with involuntary twitches of their muscles triggered by the eel's exploratory shocks. Their twitches send ripples through the water, which the eel detects with its pressure-sensitive lateral line. It then unleashes a volley of shocks like a Taser gun that overstimulates the victim's nerves, causing its muscles to contract and putting it temporarily out of action.

Catania has also worked out what may have happened to Alexander von Humboldt's horses, 200 years ago in the swamps of Venezuela. On his journey around South America, Humboldt asked local fishermen if they could obtain some live Electric Eels for him. Their response was to drive horses into a pond and watch as the eels launched a vicious attack. The men shrieked at the horses and stopped them from running away; two horses drowned and several more staggered off and collapsed. This was, Catania thinks, the eels standing their ground as they do in a situation that naturally arises every year. During the rainy season, water from the Amazon and Orinoco Rivers floods into surrounding rainforests and savannas, and fish move into a temporary wetland. Later, when the rains stop and the waters recede, the fish become stranded in isolated pools; this is when Humboldt was there, during the dry season. Electric Eels are accustomed and well adapted to this – they breathe air and can survive as ponds grow stagnant. But isolation makes them vulnerable, and predators are drawn to these ponds filled with fish that can't easily escape. Nevertheless, the stranded eels have an effective way of fighting back.

While doing other experiments, Catania saw his captive Electric Eels attack the net he used to scoop them from their tanks. The eels repeatedly charged at the net and leapt from the water, firing electric shocks into the metal handle. To measure the scale of these attacks, Catania lowered a metal pole into the tank that was hooked up to a voltmeter (the eels frequently fired 200-volt shocks). He even set up up a life-sized model of a crocodile head covered in LEDs that lit up whenever an eel shocked it. The eels deliver these powerful shocks by jumping from the water and short-circuiting their electric organ directly through another animal's body. This is more potent than if the eel simply fired the charges into the water in which the target animal is standing or swimming.

In his latest (2017) study, Catania found out first-hand just how potent the shocks can be. He rigged an experimental set-up to measure the current that flows through a living human arm when an eel attacks, using himself as the test subject. The chosen eel was relatively small, a 40cm (16in) juvenile, but nevertheless it could deliver shocks that peaked at 50 milliamps, 'greatly exceeding thresholds for nociceptor activation', as Catania's paper reports. In other words, it was dreadfully painful. Nevertheless, his arm didn't go stiff; his muscles weren't being over-stimulated and locked rigid. Rather than incapacitating a victim, he thinks the purpose of the eels' leaping attacks is to deter potential predators by giving them a sharp, painful shock.

Catania is confident these leaping eels are not hunting. They don't bite and chew their food, and they couldn't swallow anything as big as a crocodile, a horse or a human. Instead, he thinks the small aquarium tanks in his lab may convince the eels they've been trapped in a shrinking pond and are in danger from predators, just as they are during the dry season in their native habitat. In this situation, when something big and threatening looms close, the eel's innate response is to defend itself and make sure the intruder gets a lot more than it bargained for.

Charles Darwin was well aware that waters around the world are inhabited by various electric fish. In his book *On the Origin of Species*, he pondered how they evolved. '... if the electric organs had been inherited from one ancient progenitor ... we might have expected that all electric fishes would have been specially related to each other.' But, as Darwin knew, the electric fish aren't all close relatives. Sleeper rays, numbfish, Coffin Rays and torpedo rays are all electric elasmobranchs. Across in distant parts of the fish evolutionary tree, among the teleosts, there are electric stargazers, catfish, elephantfish and knifefish. Darwin regarded them all as important examples of a phenomenon now known as convergent evolution, although he didn't use that term. This happens when distantly related species evolve to look or behave or somehow operate in a similar way. As Darwin wrote, 'I am inclined to believe that in nearly the same way as two men have sometimes independently hit on the very same invention, so natural selection ... has sometimes modified in very nearly the same manner two parts in two organic beings.'

This is how tree-climbing primates from Madagascar called Aye-ayes and marsupial possums in Australia came to exploit the same food source as woodpeckers. All three groups dig holes in trees and pull out grubs from beneath the bark; the birds use their beaks and long tongues, while Aye-ayes and possums have buckteeth and long fingers.*

---

* Malagasy folk stories tell of Aye-ayes poking fingers into people's ears while they're sleeping and picking out their brains.

Similarly, fish evolved to be electric on at least six separate occasions. Darwin would have been astounded to know how this happened, and how evolution can repeatedly arrive at the same outcomes. Electric organs are packed with electrocytes made from modified muscle fibres, co-opted from different parts of the fish's bodies. In electric rays these cells are revamped gill muscles, forming two kidney shapes on either side of their round bodies. When hunting, the rays wrap up prey in their wide pectoral fins and zap them with electric shocks. Northern Stargazers live along America's eastern seaboard, buried in sand and leaving just their eyes poking out. Modified eye muscles generate weak electric shocks that may confuse their prey or deter approaching predators. As for those powerful Electric Eels, more than three-quarters of their bodies are made up of thousands of electrocytes, derived from muscles all along their bodies.

Molecular studies are revealing that all these fish use the same genetic toolkit to become electric. They all follow the same developmental pathways and switch the same sets of genes on and off. The process is complex, involving muscle cells getting bigger, losing their ability to contract and instead pushing lots of ions across cell membranes, creating a flow of charge. Despite evolving millions of years apart, in the oceans and in freshwaters and in different regions of the body, all the fish's electric organs evolved in essentially the same way.

Mysteries still remain, including how electric fish manage not to electrocute themselves while they're hunting. It might be that they insulate their vital organs under layers of fat, or they have well-insulated nerve endings, but for now nobody's quite sure.

What goes in

Bumphead Parrotfish live up to their name very well. They have bumps on their heads and teeth fused together into sharp beaks. They use their stocky foreheads in ritualised head-butting contests that were caught on camera for the first time in 2012. Like Humphead Wrasse (whose humps, as far as we know, are purely for show and not for fighting), these enormous parrotfish congregate to spawn, and it's then that fully grown males compete for supremacy by charging at each other and banging heads – with a loud crunch. The rest of the time these parrotfish are also quite noisy, taking great rasping bites of coral.

In Palmyra Atoll, in the middle of the Pacific, researchers swam across shallow reefs following Bumphead Parrotfish and watching them for hours on end, carefully noting down every bite they took. This was an impressive feat of fish-watching. The methods noted in the published study specify that only observations lasting 60 minutes or longer were included in the results. The longest the snorkellers swam non-stop alongside a fish was 5 hours 20 minutes. On average, each parrotfish took three bites of coral every minute.

Like elephants on a savanna, these are beasts that leave a clear mark on the world around them. When the parrotfish bite off mouthfuls of coral, they also dislodge smaller coral fragments, some of which survive and stick themselves back to the reef, growing into new colonies. And the parrotfish's bite-marks clear space for new coral larvae to settle. They excavate vast amounts of living and dead coral, reconstituting and redistributing limestone and sediments, and they play a major role in the dynamic rhythms of the reef.

Adding up their feeding habits over the course of a year, a single, mature parrotfish eats between four and six tonnes of solid limestone reef. They need to eat this much because there's not much nutrition to be had in this coral-based diet. Food for the fish is the thin film of living tissue that covers the calcium carbonate skeleton, and out of everything they swallow they absorb only around two per cent. A second set of teeth at the back of the throat – known as the pharyngeal mill – grinds the coral down to a powder. Nutrients are absorbed through their long intestines, and whatever's left over comes out the other end.

Fish faeces and urine provide important nutrients for other organisms. In the early 1980s, Judy Meyer from the University of Georgia watched fish on the Caribbean reefs of St Croix, in the US Virgin Islands. During the day, she saw that some thickets of branching corals had a resident shoal of fish, called grunts, with blue and yellow stripes and large, silvery eyes. At sunset, the grunts would swim off to a nearby seagrass meadow, where they filled their bellies with molluscs and crabs. By sunrise the fish were back on the reef, sheltering among the branches of their coral homes, digesting and defecating. Meyer and her team found that nutrient levels were five times higher in water near corals with fish tenants compared to those without, most likely because of the grunts' daily evacuations. Over the course of a year, she saw that the corals with fish grew twice as fast as equivalent colonies that didn't get a regular spattering. It seems these mobile vertebrates forge an important link between the two habitats, sea grasses and coral reefs, from one end of their digestive tracts to the other.

More recently, fish ecologist Jacob Allgeier spent years working out just how much coral reefs rely on fish pee. The practical side of his work, with collaborator Craig Layman at North Carolina State University, involved catching hundreds of fish species and carefully placing them one by one in plastic bags of seawater, for half an hour at a time. By measuring the water's nutrient content before and after, they calculated how much phosphorus and nitrogen each fish released, chiefly in their urine but also leaking across their gills. Allgeier combined this information with data from his colleagues, Abel Valdivia and Courtney Cox, who had the enviable task of surveying fish populations on hundreds of coral reefs across the Caribbean, some heavily fished and others highly protected where virtually no fishing took place. Crunching all the data, Allgeier estimated that when reefs are depleted of their fish, they're starved of up to half the nutrients available in healthier, fish-rich, urine-rich habitats.

The nutrient balance on coral reefs is precarious. They can easily have too much or too little. Reefs evolved to thrive in clear, nutrient-poor tropical waters; these ecosystems are highly efficient, recycling and reusing the limited nutrients (in contrast, there are nutrient-hungry ecosystems like kelp forests that require constant feeding, often from nutrient-rich waters welling up from the deep). It's well known that nutrient pollution is a problem for reefs, when sewage and farmland runoff pour phosphates and nitrates into coastal waters, causing seaweeds to smother corals and take over reefs. But there's a flip side, too. Reefs are also worse off if they lose their natural source of nutrients.

Studies like Allgeier's are showing that much of a reef's crucial nutrient pool is locked up in fish, especially big fish – the biggest produce the most faeces and urine. He didn't wrestle a metre-long (3ft) Bumphead Parrotfish into a plastic bag. If he had, he would have needed a very large bag, not only for the fish but also for its droppings. Several times when I was diving in Palau, Bumpheads emptied their bowels in front of me, sending a white, chalky plume trickling through the water. Those Bumphead-watching researchers on Palmyra Atoll noted that the adult fish defecate more than 20 times an hour.

Fish make other important contributions to the world around them. Many of the Bumphead's relatives, like Dusky, Ember and Daisy Parrotfish, are vegetarians that have strong jaws to crop seaweeds from reefs; at the same time they scrape up chunks of limestone rock. All of this passes through the parrotfish and comes out the other end in an altered state. In 2015, Chris Perry from the University of Exeter in the UK led a research team to the Maldives, to analyse the origins of the sands that build these low-lying islands. The team found that parrotfish are responsible for making more than 85 per cent of the sand-grade sediment in and around Vakkru Island; in other words, parrotfish built most of the island with their poo. So should you ever find yourself strolling along a tropical beach, spare a thought for the fish that are industriously feeding, chewing and defecating, down beneath the waves, helping to produce the gleaming white sand between your toes.

Vatnagedda

Iceland, 16th century

Iceland has very few animals living on land, but the rivers and lakes and surrounding seas are awash with mysterious fish. Some are harmless creatures, and people use them for their own good. Five eels drowned in liquor will protect whoever drinks it from becoming intoxicated; stones extracted from inside a skate's head can make a person invisible, although only for an hour at a time. But many of the Icelandic fish are dangerous and best avoided. The *lodsilungur* is a trout that grows a furry white coat to keep warm and has poisonous flesh. The *ofug-nggi* looks like a trout with coal-black skin, and it swims backwards. Eat one of these fish and the result is instant death. Should you ever go near to a ditch or a pool of stagnant water, watch out for the *hrokkull*. A wizard made this fish by taking a dead, half-rotten eel and bringing it back to life. Step in water where a *hrokkull* lives and it will coil around your leg. It has venom so strong it dissolves skin and bones, and unless the fish is quickly unwound, your leg will be cut right off. And the most venomous fish of all in Iceland is the *vatnagedda*. Shaped like a small flounder and flaming gold in colour, it is very rare indeed and is only ever seen on a foggy night before a violent storm. To catch one, bait your hook with gold and wear gloves made of human skin. Then keep it in a glass bottle, wrapped in layers of horse skin, otherwise it will burn through and sink down into the earth. This fish protects against evil spirits, and can ward off even the most powerful ghosts.

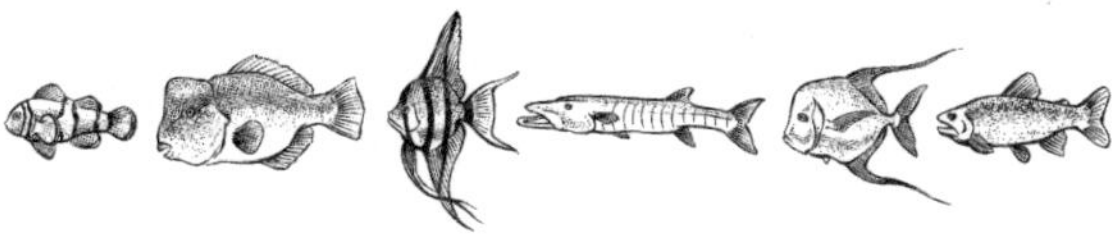

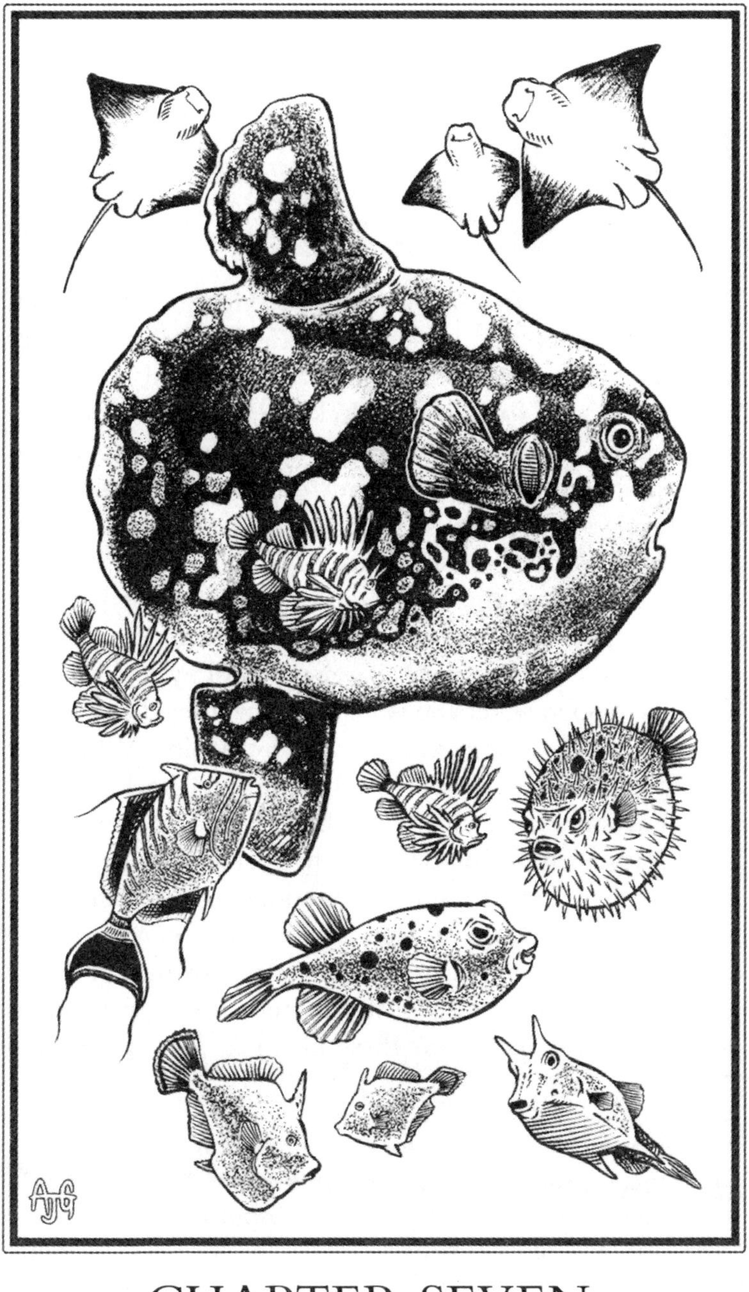

CHAPTER SEVEN:
TOXIC FISH

CHAPTER SEVEN

Toxic fish

In the early 1970s, marine biologist George Losey from the University of Hawaii spent 250 hours underwater watching little fish called Eyelash Harptail Blennies: 'eyelash' because they have a black line running back from each eye and look like they're wearing a thick mascara; 'harptail' because their yellow tails are roughly harp-shaped, with fin rays in place of strings. Losey was scuba-diving in the central Pacific, at Enewetak Atoll,* and he was interested to know how these fish behave in the presence of large predators, and how predators respond to them.

Losey took on the role of a large predator by approaching blennies in open water and watching what they did. Usually the fish initially swam slowly away from him. Whenever he stopped, the fish turned around and came back by short stops and starts until they were hovering right in front of him. Blennies living in holes in the reef would leave their shelters when Losey approached, come right up and stare him in the face. The fish were all a fraction of his size, up to 11cm (4.5in) long, and yet they seemed fearless of him.

In experimental tanks back on land, Losey watched as predatory fish tried to eat these blennies. On swallowing

---

* Along with Bikini Atoll to the east, this is where the US government detonated 67 nuclear bombs in the 1950s and 60s (it was this radiation that scientists used to estimate that Greenland Sharks may live for 400 years or more). In the 1970s, on Runit island in the Enewetak Atoll, contaminated radioactive soil and slurry were scraped up into a pile and covered in concrete panels; this was intended to be a temporary measure but it is still there today, and it's started leaking.

one of the little fish, a grouper immediately began trembling and quivering its head, and awkwardly sticking out its jaws. A few seconds later the blenny swam out of the predator's mouth, apparently unharmed.

Harptail Blennies owe their unusual confidence, in part, to their dentition. They belong to a group of fish called the fangblennies or sabretooth blennies, which all have a pair of ferocious teeth in their lower jaws. The harptails also give more than just a sharp bite, as Losey found out during his studies in Enewetak.

In his 1972 paper describing this research, Losey explains how he caught two blennies that were 'placed in a small bag in my bathing suit.' Perhaps there was nowhere else to stash the captured fish, but whatever reason he had for putting them there, he swiftly obtained some sharp personal observations on the nature of their fangs. This insight was, he wrote, 'inadvertently provided by bites in the more tender area of my hip … the bites were immediately painful, not unlike a mild bee sting.'

Losey, ever the diligent scientist, recorded the progress of his wounds, which bled for ten minutes and the inflamed, red area grew from a few millimetres after two minutes to 10cm (4in) a quarter of an hour later; the inflammation continued for four hours, and the immediate area of the puncture wound remained inflamed for the next 12 hours. 'The tissues were somewhat hardened for several days,' he noted. Losey saw for himself that this particular fangblenny species is undoubtedly venomous.

Fish, it turns out, are the most venomous of all the vertebrates. Up until 10 years ago, it was generally thought only around 200 fish species have venom. A more thorough look at the matter, however, recently revealed that there are closer to 3,000 species of fish that you definitely don't want to put inside your bathing suit.

Venoms are yet another facet to the fish's immensely successful lives, and a prime way in which they avoid becoming someone else's dinner. Like electric shocks, fish have evolved venoms repeatedly in different groups – at least 18 times. There are venomous catfish and chimaeras, hornsharks and stingrays, rabbitfish and surgeonfish. Based on species number, you're far more likely to be impaled by a toxic fish than bitten by a snake or scratched by a platypus's venomous claw.

The good news is that fish venoms are unlikely to kill you. The bad news is they are among the most excruciating stings of any venomous creatures. With possibly one exception – the one-jawed eels, which we know very little about – fish don't use their chemical weapons to attack but to defend themselves, and predators quickly learn to avoid them. When Losey put those fangblennies in his shorts, they were scared and knew they were in trouble; they used their hollow teeth to deliver a cocktail of chemical deterrents.* A 2017 study investigating the venom composition of this particular species found it to contain, among other things, opiate peptides that bind to the same nerve receptors as heroin and morphine. The main effect of this venom is to cause blood pressure to crash by up to 40 per cent. If your blood pressure dropped that low, you would certainly feel dizzy and you might need a bit of a sit down. Likewise, the venom probably confuses a predatory fish and sends it into the kind of woozy state that makes it easier for a small fish to climb safely back out of its mouth, just as Losey observed in his studies.

* Recent studies suggest that the fangblennies' ancestors first evolved huge fangs to take bites out of other, larger fish. It was only later in the group's evolutionary history that some of these fish, including the Eyelash Harptail Fangblenny, evolved venom and a deep groove in their teeth that it flows along, like a hypodermic needle. It's uncommon for venomous animals to evolve this way round (i.e. teeth first). Snakes evolved venoms first and presumably dribbled on their victims; subsequently, they evolved hollow fangs as a more effective means of administering toxins.

In general, if you steer clear of venomous fish they will leave you alone, too. Some, such as the lionfish, use bright colours to warn of their stings and are quite easy to spot. However, plenty of venom-loaded fish are well camouflaged, and live hunkered on the seabed or riverbed. Beachgoers in Britain occasionally tread on weaver fish, which hide in the sand. Like most venomous fish, they inject their toxins through modified fin spines. Similar agonising injuries occur on the US coasts where stargazers live. And if you step on a stingray it will flick its tail up and impale your leg with a venomous barb.*

The most dangerous venomous fish are probably the stonefish, a family whose members disguise themselves as weedy rocks. Even if you know there's one there, they're almost impossible to spot. One in particular, the Rough Stonefish, also known as the Warty Ghoul, has a row of 13 spines along its back (in 1766 Carl Linnaeus appropriately named it *Synanceia horrida*). Every year, hundreds of people in Australia accidentally step on stonefish; the person's weight squeezes venom ducts at the base of the spines that fire deep into their feet. This triggers a crippling pain that can last for days, and while there is an anti-venom it's a good idea to look carefully where you're treading, and not to touch anything on a coral reef because it's easy to be fooled by a stonefish's flawless disguise.

Another group of fish is notorious not because they sting but because they're poisonous – eat one and it could easily kill you. For centuries, people have been deeply intrigued by pufferfish, from the Ancient Egyptians who carved pufferfish hieroglyphs to the Japanese diners who still pay huge sums and risk their lives to eat them. Until regulations

* If you're wading through water where there might be stingrays, shuffle your feet along so you don't tread on one but instead encourage it to swim away from your footsteps. And if you are stung by any fish, the best treatment is hot water (but not scalding) to denature and deactivate the protein-based venom.

were enforced requiring chefs to train for years and obtain a pufferfish-handling licence, dozens of people died every year in Japan from *fugu* poisoning.* Now the annual death toll is down to two or three unlucky (or foolhardy) diners.

The reason puffers are so dangerous to eat is a poisonous alkaloid called TTX, short for tetrodotoxin. It accumulates in pufferfish livers, in their reproductive organs, skin and intestines, all the bits that skilled chefs know how to slice out. A single milligram of this potent neurotoxin – a droplet small enough to sit on the head of a pin – will kill a fully grown human. Cooking doesn't deactivate it. There's no known antidote.

Pufferfish don't produce tetrodotoxin themselves, but they get it from food containing TTX-making bacteria. Give them bacteria-free food and puffers gradually lose their potency. In this way, fish-farmers have produced puffers that are guaranteed safe to eat, but they're proving unpopular among Japanese diners who still value the thrill of eating wild-caught *fugu*.

You should avoid eating various others animals that also contain TTX. In New Zealand in 2009, five dogs died after eating sea slugs that had washed up on a beach. Pester a blue-ringed octopus and you're likely to come to a swift end thanks to a tiny, often painless bite that's lethally infused with TTX. This chemical could even add a scrap of truth to magic potions containing 'eye of newt': drop a Japanese Fire-Belly Newt, a Harlequin Toad or a Pumpkin Toadlet into a cauldron and it will make for a deadly TTX-laced brew.

Exactly how newts, octopuses, sea slugs, puffers and the rest of them don't poison themselves is a puzzle that's recently been solved. TTX acts by binding to sodium channels in nerve cells and stopping them from sending signals. This silences transmissions between the nervous system and muscles, and ultimately causes paralysis (and

---

* *Fugu*, meaning 'river pig', is the name of the Japanese delicacy made from fish in the genus *Takifugu*.

often death by suffocation). It turns out that it's a fairly simple process to halt the effect of TTX. All that's needed is a genetic mutation altering a few amino-acid building blocks that make up these protein-based sodium channels. This stops TTX from binding and blocking signals, so even if there's poison around, nerves will work as normal, resulting in an animal that's immune to TTX. Such resistance has evolved repeatedly among the pufferfish, with exactly the same genetic mutations switching out the same amino acids in the channel proteins each time. Within narrow constraints, in this case of boosting resistance to a poison while keeping nerves working, natural selection can be highly predictable, going back time and again to tweak the same genes in similar ways. In California, there are snakes that happily scoff poisonous, TTX-tainted newts thanks to this same mutation in their nerve-channel proteins. They are so resistant to TTX that the lethal dose for a single snake is enough to kill 600 people.

Resistance to this poison provides pufferfish with multiple benefits. It expands their diet, allowing them to feed on things contaminated with TTX, and it gives them a potent chemical defence. Male puffers have even evolved something of a fondness for the stuff. Females smear it on their eggs to discourage predators from nibbling them, and males are attracted to the smell.

Besides loading their tissues with TTX, pufferfish have another unusual strategy for defending themselves. When calm and relaxed they have lumpy, un-streamlined bodies, a wide, pouting mouth and bulging eyes. Get one angry or scared, however, and a puffer will inflate itself into a taut, prickly ball. Try swallowing one of those.

Puffers belong to an order of fish formerly known as plectognaths,* but they are now usually referred to by a bigger mouthful, the tetraodontiformes, so named because

* From Ancient Greek words *plektos* meaning 'twisted', and *gnathos* meaning 'jaws'.

many of them have four buckteeth;* it's also where tetrodotoxin gets its name. Among the puffers' relatives are various other well-defended fish. Porcupinefish also inflate themselves and have elongated scales with three-pronged bases, which lock upright and protect the inflated fish inside a nibble-proof cage of spines. Boxfish, cowfish and trunkfish all come encased inside a rigid bony box, triangular or square in cross section, and fashioned from large, hexagonal scales (cowfish are so named because they have a prong above each eye, like horns). As well as its suit of armour, a scared or stressed boxfish can ooze a poisonous goo that wafts through the water, driving off unwanted intruders. A triggerfish evades capture by darting into a hole in a reef and flicking up a sharp spine on its back – the trigger part of its name – wedging itself firmly in place so a predator can't pull it out. And sunfish, another tetraodontiform and the biggest of all the teleosts, generally avoid predators simply by being incredibly big. They start life as tiny larvae but grow very fast, gaining up to a kilogram (2lb) in weight every day; the biggest sunfish on record weighed 2.3 tonnes (5,000lb), the same as a fully grown female African Elephant.

The puffers and their formidable kin have attracted much attention from researchers, who've tried to understand and harness their powers. One woman in particular devoted much of her life to studying these fish, and to uncovering the secrets of their puffs and poisons.

The lady and the puffers

Eugenie Clark is most fondly remembered as the Shark Lady. In the 1940s, she embarked on a trailblazing career at a time when few women were involved in science, let alone setting off on their own as solo explorers. She was the first scientist to discover that sharks aren't mindless killing machines but that they can learn and remember things, and

* *Tetra* meaning 'four', and *odont* meaning 'tooth'.

are just as intelligent as many other vertebrates. But sharks weren't the only fish she studied. She could just as well be known as the Pufferfish Lady.

I had the pleasure of meeting Genie in 2011. Mote Marine Laboratories in Florida had invited me to give a public lecture on Valentine's Day on the subject of seahorses and their unusual sex lives. I immediately made enquiries as to whether Genie might be in town during my visit. In her retirement she often returned to Mote, the laboratories she founded in 1955. Her assistant emailed to say we could have lunch together the day after my speech.

To my surprise, though, I first met Genie on that Valentine's night. I spoke about seahorses for my allotted hour, answered questions from the audience, then sat at a little table to sign copies of my book. I looked along the growing queue and spotted someone I recognised standing patiently in line. A few moments later a lady stepped up, leant in and whispered to me, 'This is my friend Eugenie Clark, she'd like to have her picture taken with you.' I stared, bewildered for a heartbeat, then grinned back and awkwardly replied, 'I know who she is.'

That picture shows a smiling Genie wearing a sweatshirt with two orcas leaping across the front, her arm wrapped around my shoulder giving me a gentle squeeze, and I am still grinning.

The next day over lunch, my nervous excitement eased and it felt like Genie and I were just two friends catching up. She seemed to be as interested in me and my work as I was in her. With a sparkle in her eyes, she asked me about the places I'd been and parts of the ocean I'd seen.

Up until that point, the Genie I knew mostly came from her books. When I met her she was shortly due to celebrate her ninetieth birthday. Stretching behind her was a tireless and eminent career of research and adventures, spanning 70 years and showing no sign of letting up.

Born in 1922, Eugenie Clark was raised in New York City by her Japanese mother, after the death of her father when she was two. The first fish she saw were at an aquarium in Battery Park, on the southern tip of Manhattan overlooking the Statue of Liberty. It was a Saturday, and on the way to work her mother dropped off the nine-year-old Genie at the aquarium to keep her occupied for a few hours. 'So casually, so by chance, I entered the world of water,' Genie wrote in 1953 in her first book, *Lady with a Spear.* 'Leaning over the brass railing, I brought my face as close as possible to the glass and pretended I was walking on the bottom of the sea.'

Genie went back to the aquarium every weekend after that, and quickly caught on to the idea of keeping fish for herself. She persuaded her mother to make space in their small apartment for an aquarium tank. Her collections expanded to include other animals – salamanders, snakes and toads – and she began bringing home dead cats and monkeys from the local pet shop to dissect. Always, though, it was fish that kept her gazing and wondering. 'Throughout high school,' she wrote, 'fish were on my mind.'

Genie majored in zoology at Hunter College in Manhattan's Upper East Side, and when she graduated she dreamed of getting a job like her hero, the world-famous deep-sea explorer, William Beebe,* from the New York Zoological Society. In the 1930s he climbed inside a small metal sphere, together with its inventor Otis Barton, and the pair descended more than 900m (3,000ft) into the sea off Bermuda. Beebe and Barton set a string of records for the deepest-diving humans and they were the first people to see deep-sea animals alive in their native habitat.

By the time Genie finished her undergraduate studies World War II had broken out, and there were few opportunities for young American zoologists. Her mother advised her to take courses in typing and shorthand in case

* Pleasingly, pronounced 'bee-bee'.

she could get her foot in the door as some famous ichthyologist's secretary.* Genie didn't take her mother's advice. She turned her hand instead to chemistry, and got herself a job researching industrial plastics to pay her way through grad school. In the evenings she took classes at New York University, including her favourite – ichthyology. Her professor, Charles Breder, was curator of fish at the American Museum of Natural History, where he introduced her to the fish that would keep swimming through her life for many years to come.

In the Museum's Hall of Fishes, Genie saw tetraodontiformes (or plectognaths as they were then still known) for the first time. She peered into glass cases at the dried, pickled bodies of sunfish, pufferfish, triggerfish and boxfish, and under Breder's tutorage she began to study them in detail. For her first published paper in 1947, Genie co-wrote a 33-page treatise with him on these fish. They compiled an evolutionary tree, studied how the young develop from embryonic balls of cells to wriggling larvae, and investigated how some inflate themselves. She saw that many have big, stretchy extensions in their stomachs. Blowing air into their guts she investigated which parts were the stretchiest and might have puffed up in life. Many of the puffers and porcupinefish obviously had inflatable stomach sacs, but the giant sunfish she looked at did not.

It had long been thought that when puffers are alarmed, they swim upwards, poke their lips above the waves and suck in air, then float like a beach ball on the surface beyond the reach of their aquatic enemies. Sure enough, if you haul a puffer from the water, as many fishers and scientists have done, it will indeed gulp air and inflate. But as Genie knew, puffers in their natural environment don't go to the trouble of swimming to the surface – they simply suck in the water

* As did Lotte Baierl, who had been assistant to Hans Hass, an Austrian underwater explorer, before marrying him and becoming co-presenter on his documentaries.

around them. It takes roughly 15 seconds and 40 or so puffs for a fish to inflate up to three times its original volume. To accommodate such massive expansion, puffers have lost their ribs and evolved incredibly stretchy skin, eight times stretchier than normal fish skin. And those elastic stomach sacs can hold a lot of water. Twenty years previously, Breder had caught dozens of puffers from the lower New York harbour. He gently prodded the living fish so they inflated, then persuaded them to spit out their water into a measuring jug. The results of this are presented in their paper: a medium-sized fish, about 20cm (8in) long, sucked in more than a litre (1¾ pt) of water, enough to fill five average water balloons.

On finishing her master's degree in 1946, Genie moved to California to continue her studies at Scripps Institute of Oceanography. A year later, and still only 25 years old, she was offered her first overseas job as a marine biologist. It seemed that her dream of following in Beebe's footsteps was going to come true after all. The US Fish and Wildlife Service was interested in developing new fishing grounds in the Philippines, and Genie was recruited to carry out fish surveys around the islands. But she never made it that far. During a stopover in Hawaii she was held up, and told that the FBI was checking on her Japanese ancestry. After two weeks of waiting she handed in her resignation, convinced she'd been pushed out for being the only female scientist on the programme. As she wrote in *Lady with a Spear*, 'They hired a man in my place.'

She didn't give up, and soon had another chance to explore tropical waters. Returning to New York she continued her PhD, focusing on the sex lives of some of the freshwater fish species, swordtails and platys, that she'd originally kept as pets. By then, Charles Breder was director of the Lerner Marine Laboratory in Bimini, in the Bahamas. Genie spent several months there, and for the first time got to work not with stiff, formalin-soaked specimens but with living fish.

Using nets, traps and hooks, Genie caught hundreds of live plectognaths from the waters around Bimini and kept them in ocean pens and concrete tanks in the lab. Watching them for hours, she worked out why some of her fish did headstands. William Beebe had been the first to point out the unusual habits of the Fringed Filefish (another plectognath) of Bimini. Chiefly it's the males that put on splendid displays, opening out a large flap of skin on their bellies and sticking out all their fins. The male dips his nose downwards and vigorously vibrates his body. This usually happens in confrontations between two males. Both begin to flare out their fins and lower their head, but only one – usually the bigger of the pair – completes the full, flamboyant headstand. The subordinate fish folds away his fins and retreats.

Genie discovered a strict hierarchy among male filefish. One male in her study was clearly the boss, and won every headstand contest he took part in. The second-in-command won every time except against the boss. 'And so on down the line,' wrote Genie. The lowest-ranking fish soon died because all the others beat it to the food at feeding time. 'My poor sickly fish who died wasn't exactly "henpecked",' she wrote, 'but he was "headstood" by all the others.'

As she was nearing the end of her doctoral studies, an opportunity came Genie's way to explore even more distant seas. At the end of the war, many Pacific islands had come under American control. The US Office of Naval Research was keen to learn more about these far-flung outposts, and put out a call to interested researchers. Genie applied and, despite the concerns of others that such remote work might not be suitable for a single woman, she was given two weeks' notice to pack her equipment. Her mission was to investigate fish poisoning, a problem that's especially common in tropical waters and was a pressing issue among members of the American armed forces stationed across the Pacific. Eat a toxic fish, a fresh one that was noxious from the get go and hasn't just gone bad from decomposition, and you're likely to experience a range of symptoms. Within the next day or so you may be hit by bouts of vomiting and diarrhoea, stomach pains, convulsions and paralysis, along with some more peculiar sensations, such as feeling hot when it's cold and *vice versa*, and becoming quite convinced that your teeth are about to fall out. Besides pufferfish TTX there are other fish-borne toxins to watch out for; there's ciguatera, saxitoxin, plus other, unidentified chemicals that can make you hallucinate. By gathering fish and sending samples back to labs for chemical analysis, Genie's work would help find out which species were safe to eat. In June 1949, she boarded a military seaplane in California, and with the roar of four propellers in her ears she headed towards the setting sun to spend four months island-hopping in search of poisonous fish.

Her first scheduled stop was the island of Guam, where she immediately started tapping into the knowledge and skills of local fishermen. One man she met had a large fish-trap made of chicken wire and bamboo, with seven good-sized pufferfish inside. 'I pointed to them enthusiastically,' she wrote, 'but the fisherman shook his head, made a motion of eating, then held his stomach, and made painful grimaces.' These were precisely the sorts of fish Genie was looking for.

Further east, Genie visited the remote islands of the Palau archipelago, travelling around on local ferries and copra boats. She stayed with small fishing communities, where she learned to spear-fish, and local women danced for her and taught her to chew betel nut without making a mess on the floor. The men helped her find fish, based on pictures she drew in the sand. Everywhere she went, Genie kept an ear out for tales of poisonous fish. One concerned a rabbitfish, known locally as *meas*, which she had eaten many times without getting sick. Rumour had it that in one village on Babeldaob, the biggest island in Palau, the *meas* were dangerous to eat. Genie went to investigate, accompanied by a local spear-fishing champion who went out at night and skewered the rabbitfish while they dozed in shallow seagrass meadows. The villagers declared the fish were perfectly safe. Between October and January, they said, eating *meas* will make you sleepy, angry or laugh your head off. This is when the wind blows steadily from the east and a particular green seaweed grows in the bay. It's plausible that rabbitfish consume some noxious substance in the seaweed, which builds up inside them and makes them seasonally intoxicating. But it was August. Genie had come too early to get tipsy on rabbitfish. She tried some anyway, tasting slivers of raw *meas,* but didn't get so much as a headache.

The people of Palau aren't the only ones who know about the inebriating effects of certain fish. In the Mediterranean lives a sea bream that goes by various names, one of which is the Dreamfish* – a name they occasionally live up to. In 1994, a man holidaying in Cannes on the French Côte d'Azur was admitted to hospital after he thought frenzied animals were shrieking at him, and he saw giant insects crawling all over his car. A day later, he had fully recovered from his Dreamfish dinner. Also on the French Mediterranean coast, in 2004, an elderly man cooked himself a Dreamfish, and two hours later was tormented by human screams and wailing birds and for the next two nights had terrifying nightmares. There are even stories that Ancient Romans used these trippy fish as recreational drugs. But what if the mind-bending properties of a toxic fish were used to deliberately render someone catatonic, perhaps for months or years at a time? Could this terrifying prospect be a reality?

Pufferfish and the living dead

In the 1980s, dried, powdered extracts of Caribbean pufferfish were stirred into a storm of controversy surrounding the myths and realities of zombies. It was in the early part of the 20th century that western culture first

---

* Biologists would call it *Sarpa salpa.*

really took an interest in these legends (at which time US forces were occupying Haiti). Corrupted versions of the Haitian vodoun religion, which combines elements of West African magic and Roman Catholic rituals, were misspelt and misunderstood as voodoo. They came with ideas of sticking needles into wax dolls of your enemies, and the possibility that dead people can wake up and stagger about, unleashing untold troubles.

In Haiti the threat of zombis (the original Haitian spelling, without an e) is considered to be very real. Many children are supposedly scared not of the zombis themselves, but of being turned into one. They're brought up to believe that breaking the rules of the secret vodoun societies is punished by zombification. A priest transforms the miscreant into the undead, keeping their soul in a jar and raising them from the grave as a slave with no will of their own. Making zombis is illegal under state-sanctioned law. The act of convincing someone they've died and come back to life as a zombi is deemed to be attempted murder. Bury someone alive and that counts as *actual* murder, whether or not the victim survives.

In 1982, a PhD student from Harvard University travelled to Haiti to find out how to make a zombi. Wade Davis intended to procure the potions that vodoun priests allegedly used to zombify people. His advisors at Harvard were convinced these concoctions could transform modern medicine and surgery. Imagine being able to put someone into an oblivious near-coma, then wake them up when you want. Even researchers at NASA were interested. These zombi potions could, perhaps, keep astronauts in suspended animation on long missions across the galaxy.

Propelled by this strange collision of folklore and science fiction, Davis spent several months in Haiti and came back with eight samples of zombi potion. He had also planned to commission a zombi and watch the priests at work. No doubt to the immense relief of Harvard's ethics committee, that never happened. But even without becoming an accomplice to attempted murder, his work unleashed a scandal that went on for years.

Davis boldly announced that he had uncovered the secret of zombis. He reported that to trick someone into believing they've died then come back to life as a perpetual slave, vodoun priests administer a heady brew of plant and animal extracts, including frog, centipede, tarantula and human remains. This causes sub-lethal poisoning and the appearance of death. Other concoctions then keep the victim in a permanent, zombi-like state. The key ingredient in the death potion, according to Davis, was tetrodotoxin taken from pufferfish.

An almighty slanging match immediately exploded as academics on all sides tore his arguments to shreds. Ethnographers were appalled by Davis's methods. He'd spent far too little time in Haiti, they said. He interviewed only a handful of people, including one who claimed to be a former zombi, but Davis spoke no Creole and so crucial details may have been lost in translation. He failed to prove any genuine link between the secret vodoun societies and zombification. And who's to say the priests didn't just see a chance to make some money selling hoax potions to the gullible foreigner?

Davis also managed to infuriate biologists. His PhD included no chemical tests, but nevertheless professed the importance of TTX in zombi potions. All he really had to go on was the word of the priests who listed several pufferfish species among the many ingredients they used. It later transpired that Davis had in fact done some tests on the potions, but found no trace of pufferfish toxin, and neglected to mention these negative results in his thesis (he eventually owned up to this, but claimed the tests had been poorly conducted, making the results unreliable). Toxicologists subsequently tested two of his eight potions, but the outcome was hardly compelling. There were insignificant traces of TTX, which led to no signs of intoxication when injected into mice.

To his credit, Davis made it clear that he thought the potions would probably only work on victims who believed in zombification.* He went on to clarify that the priests don't work to a strict recipe, and the amount of TTX naturally varies. Some potions are too weak and don't work, some are too strong and kill the victim outright, and others, like some gruesome retelling of Goldilocks, are just right for making a zombi.

Still, though, Davis had virtually nothing to back up his central theory that the key ingredient in all this is pufferfish poison. Instead he flipped the question on its head and demanded sceptics disprove his claims, rather than offering any real evidence.

It all got very messy, and details grew blurred in the media fracas that accompanied these academic squabbles. Davis turned his PhD into a bestselling book, *The Serpent and the Rainbow,* which then hit the silver screen. The Hollywood version, directed by Wes Craven in a follow-up to his 1984 blockbuster *A Nighmare on Elm Street*, saw Davis's character himself being buried alive and turned into a zombi. Davis publicly disowned the film, calling it 'one of the worst Hollywood movies in history'.

Wade Davis didn't pursue his interest in zombis any further, and he moved on without leaving any reliable clues as to the real powers of pufferfish. It's conceivable that Haitian vodoun priests do indeed make a potion from dried, ground-up pufferfish and use it to make slaves for themselves. People do plenty of other weird things with animals, like making magic charms from pangolin scales or eating tiger bones because they think it will make them good in bed; it doesn't mean that any of it actually works.

A far more reliable strategy would be to employ TTX as a straightforward murder weapon. At the end of Ian Fleming's fifth James Bond novel, *From Russia with Love*, our hero collapses after soviet agent Rosa Klebb stabs him

* And that presumably didn't include the lab mice.

with a TTX-smeared spike hidden in her shoe. Luckily Bond pulls through, as he always does.

In the real world, in 2011, a British man visiting Sierra Leone was probably killed by pufferfish poison, which may have been administered deliberately. TTX was found in his body after he mysteriously dropped dead a few days after a lunch meeting with a business associate; at the inquest the coroner recorded an open verdict, and refused to rule out the possibility of foul play. And in 2012, a man from Chicago, Illinois, was jailed for seven and a half years after posing as a scientist and buying purified extracts of pufferfish from a chemical supplies company. According to the *Chicago Tribune*, the convicted TTX-hoarder had been plotting to murder his wife and cash in on her life insurance. If he had got as far as administering the poison, it would have certainly been effective. He had 98mg (0.003oz) of TTX, enough to kill almost a hundred people.

After Eugenie Clark finished up her research on poisonous fish in the remote islands of the Pacific she went back to America to complete her PhD. A Fulbright Scholarship then took her to Egypt for a year and filled the last few chapters of *Lady with the Spear* with more explorations in search of venomous fish, this time in the Red Sea. Two wealthy benefactors read her book and decided to fund a new American research station, just like the one she'd worked at in Egypt. And they wanted Genie at the helm.

In 1955 she founded the Cape Haze Marine Laboratory, originally a one-room wooden building perched on the eastern shores of the Gulf of Mexico in Florida. The station was eventually relocated an hour's drive north to Sarasota and renamed the Mote Marine Laboratory, where many years later I met Genie.

We sat and chatted, and she told me some of her favourite stories from her Cape Haze days, like the time she took a baby shark on a plane ride to Japan. She'd been invited to visit Crown Prince Akihito (now the Emperor), a fellow fish-enthusiast, and as a gift brought him a shark that had learned to ring bells. It was part of Genie's ground-breaking studies of shark cognition, showing for the first time that sharks can be taught to recognise shapes and patterns and, in return for food, press their nose on a target with a tinkling bell attached. On their flight across the Pacific to Japan the obedient shark sat in a portable water tank on the aeroplane seat next to her. 'Most people didn't know,' she told me, chuckling. 'He was such a tiny little thing. He was less than two feet (60cm) long.'

In 1968 (a year in which she appeared in the ground-breaking TV series *The Undersea World of Jacques Cousteau* as on-board shark expert), Genie left Florida and moved north to the University of Maryland. There she spent the rest of her career as a professor of ichthyology. She taught and mentored thousands of students, and whenever she could she headed overseas to continue her studies. She became a pioneering scuba-diver and trained as a deep-sea pilot, venturing far deeper than her hero William Beebe had ever been, in submersible vehicles far more sophisticated than his metal ball.

When I met her, she still hadn't hung up her dive mask and had no intention of retiring from the underwater world. Three years later, in June 2014 and aged 92, Genie led a diving expedition to the Solomon Islands, where her career-spanning search for tetraodontiforms continued. The subject of this study was the Oceanic Triggerfish, a 50cm (1.6ft) long species that looks like a small, stretched-out version of a sunfish. Genie had been diving to watch them on and off for almost 30 years, in particular to learn how they build nests.

Nest-building among triggerfish, pufferfish and their relatives remains a mysterious and little-known behaviour. In 1995, divers in the Amami Archipelago in southern Japan came across an elaborate circular shape sculpted into the sandy seabed. It was 2m (6.5ft) across, and formed by two concentric rings with spokes radiating from the centre. Similar mystifying shapes appeared sporadically around the islands, and no one could work out what, or who, had made them. Then in 2011 a team of diving scientists finally caught a small male pufferfish in the act of making one of these sand-sculptures. Subsequently watching another 10 artistic puffers, the divers traced the steps involved in making the circles: the puffer draws lines in the sand by fanning the seabed with his fins, first tracing basic circles then embellishing them with ridges, swimming inwards at different angles; next he fills the central area with a doodle of wiggly lines; and for a finishing touch he collects fragments of seashells and dead coral, and arranges them carefully around the circle. The whole endeavour takes at least a week.* Then, hopefully, a female puffer shows up, inspects his work and decides to lay her eggs in the centre of the circle before swimming off. The hard-working male hangs on for another six days, guarding the nest and the developing eggs, while his artwork gradually crumbles and gets swept away in the current.†

Triggerfish nests are less intricate. Most species gather coral rubble into a mound, which they fiercely defend from intruders, including human divers. In her studies of Oceanic Triggerfish, besides the possibility of getting shooed away by an angry fish, Genie faced the added

---

* An equivalent sand-sculpture, based on relative body size, for an average human adult would be 30 metres (100ft) across.

† It's possible the male puffers make these elaborate structures to sift the seabed and create the perfect, soft spot for eggs. The radial design may channel water towards the middle of the nest, so no matter which way the current flows it brings fine sands and fresh, oxygenated water into the central spawning area. It is also, presumably, good for catching the attention of passing female puffers.

challenge of depth; these triggers commonly build their nests on deep coral slopes between 35 and 40 metres (115–130ft) down, a depth where regular scuba-divers can't easily spend much time. Nevertheless, Genie and dozens of volunteer divers together spent more than 3,300 hours underwater watching these Oceanic Triggerfish as they made and tended to their nests. The paper based on these observations was published just before Genie died in February 2015; it features maps of the triggers' nests, descriptions of the dark patterns that spread across their faces like an eye-mask when they mate, and details of how the females, not the males, stand guard over the nests.

On her final dive, Genie went to watch nesting triggerfish down at 25 metres (84ft), an unimaginable feat for most 92-year-olds. By then, she'd already been diving beyond expectations for many years. During an interview at Mote, back in 2008, she let slip the depths she had recently visited, and immediately swore the reporter to secrecy.

'Don't tell anyone how deep I went,' she said. 'I'm not supposed to do that anymore.'

Chipfalamfula

Mozambique, traditional

Chief Makenyi had many daughters. Chichinguane was his favourite, and all her sisters were jealous of her. When they went to the river to fetch mud to plaster the village houses, the older girls ordered Chichinguane to climb down the steep, slippery riverbank and fill their buckets for them. Then they left her, knowing she could never climb back up on her own.

She called for help and heard a deep voice in the river. 'What is the matter, little one?' it said. It was Chipfalamfula, a giant fish who had power over all the water. 'Come and live in my belly, and you will never want to go anywhere else.' So Chichinguane stepped into the huge mouth and slid down into the fish's belly, which to her surprise was full of people tending fields of maize and pumpkins. They were all very kind to Chichinguane and she felt happier than she ever had before.

When her mother found out what had happened, she went to the river and called out to her daughter to come back home. 'I'm a fish now,' Chichinguane said, showing her silvery scales. 'I live in the water now.' But after seeing her mother again, Chichinguane became homesick and asked the great fish if she could leave. Chipfalamfula agreed and gave her a gift to take with her – a magic wand. She tapped the wand on her silver scales and they turned into coins, which her mother used to hold a great welcome feast.

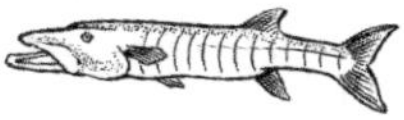

Later, when the chief's daughters were gathering firewood, the older girls sent Chichinguane and her younger sister to climb the tallest tree and cut down the highest branches.

Just then a troop of one-legged ogres came by and the other girls ran away, leaving Chichinguane and her sister trapped in the tree. The ogres saw them and began to chop the tree down, but Chichinguane used her wand to heal the axe cuts and the tree stayed tall and strong. Eventually, the ogres grew tired of chopping and fell asleep, their snores rattling through the night. The girls jumped down and ran off but the ogres awoke and chased after them. Reaching the riverbank, Chichinguane touched the water with her wand and sang, 'Chipfalamfula, shut the water off.' The river stopped flowing and the girls ran across. Then she touched the river again and sang, 'Chipfalamfula, open the water.' The ogres were only half way across when the raging torrent returned and washed them away.

On their way home, the girls came across a cave where the ogres had lived. It was full of bones from the people they had eaten, and all their gold bracelets, beads and necklaces. The girls dressed themselves in fine jewellery, then ran out into the dark forest, lighting their way with the giant fish's wand glowing brightly. Then they stepped into a clearing and saw a great palace. The palace guards saw the girls dressed in gleaming jewels, and guessed they were princesses and invited them in. The following day, Chichinguane and her sister met the king's handsome sons and accepted their offers of marriage, and lived happily ever after as princesses in the palace.

CHAPTER EIGHT:
HOW FISH USED TO BE

CHAPTER EIGHT

How fish used to be

Give me some scuba kit and a time machine and I'll set the dial to 380 million years ago, so I can watch strange fish swim through Devonian seas. This is a time when an epic evolutionary experiment is underway, trying out different ways of being a fish. Galeaspids and osteostracans, types of jawless fish, push massive head shields through the water, some bullet-shaped, some like shovels; *Doryaspis* has a rounded, armour-plated body and a prong sticking forwards like a miniature sawfish; on the seabed, snorkel tubes give away the hiding place of flattened, triangular *Eglonaspis*. Ancestors of lungfish and coelacanths prey on fidgeting clouds of worm-like fish called conodonts; acanthodians, also known as spiny sharks, glide through the water, their bodies covered in minute, diamond-shaped scales and with sharp prongs adjacent to each fin.

Most of all I want to meet the placoderms, the armoured fish that rule these seas. Tiny placoderms, small enough to fit inside a matchbox, have armoured fins sticking out stiffly to the sides – these are fish with arms, of sorts. Flattened placoderms lie on the seabed like stingrays (although these elasmobranchs didn't evolve for a while to come). And dark shadows fall through the water as the world's first vertebrate super-predators cruise around. There are at least 10 *Dunkleosteus* species, with giant, thick-set heads covered in armoured shields that extend into their jaws, forming fangs that slice past each other like gigantic, self-sharpening garden shears. Placoderms are the first fish with jaws, and *Dunkleosteus* flaunt theirs with terrifying effect. In Devonian seas, sharks are less than a metre (3ft) long and a one-tonne *Dunkleosteus,* at six metres (20ft), make easy work of swallowing them whole, tail first. For these giant placoderms, there's little to fear except others of their own kind. Titanic battles break out between these leviathans, perhaps to dispatch competitors or even for a meal. Their enormous fangs punch deep holes, slamming their jaws shut with a bite force three times more powerful than today's huge aquatic predators, the Great White Sharks and Saltwater Crocodiles. A pair of fighting *Dunkleosteus* bite again and again, until eventually the winner feels its rival's armour crack and give way.

In truth, of course, no divers will ever get a chance to watch a living placoderm, or any other ancient fish. But knowing about this lost world tells us much about how the fish we see today came to be.

Modern fish wouldn't rule the aquatic realm in the ways they do today were it not for their ancestors who thrived and survived for hundreds of millions of years. Looking into the past we can see how ancient fish adapted and repeatedly reinvented themselves while all around them life on Earth was rising and falling. They were the first to evolve many important characters that are still around today, like jaws; the articulated bones in your mouth that let you chew and grin have been vital in vertebrate evolution, allowing diverse and efficient ways of feeding. Ancient fish also experimented with a host of other features that didn't persist but were added to a catalogue of vanished oddities – all of them playing their part in the illustrious dynasty of fish.

We know so many details of these ancient lives because people have learned how to read rocks. Palaeontologists are piecing together a more detailed view than ever of ancient underwater life. They interpret fossilised bones and impressions left by bodies pressed into stone, and reconstruct the way fish used to be.

Stunning insights into the past have come from various parts of the world where exquisite fossils can be found. One such spot is a huge limestone escarpment, called the Gogo Formation, standing in a remote desert in the Kimberley, the northernmost region of Western Australia. Preserved inside Gogo are scores of animals from Australia's first Great Barrier Reef, which flourished in the Devonian period and skirted 1,400km (870 miles) along the edge of the southern supercontinent, Gondwana. When reef fish died, their bodies drifted into deep-water bays next to the reef and were swiftly buried in mud and sealed inside limestone nodules that kept the bodies intact and un-squashed. Fish fossils were first found here in the 1940s and since then teams have returned to carefully chip the nodules away. Like a palaeontologist's perfect prehistoric Easter egg, the nodules are then opened up to find the treats that lie within. In museum laboratories the nodules are gently dunked in baths of acetic acid, the same strength as vinegar, which gradually dissolves the surrounding limestone and reveals the intricate, three-dimensional fish skeletons. Not only are the fish's bones and armour plates preserved, but also their soft insides. There are individual muscle fibres and nerve cells set in stone that contracted for the last time and sent their final electrical messages roughly 380 million years ago.

There are also tiny fish inside other fish. Originally these were thought to be fossilised snapshots of predation and the larger fish's last meal. But there were no signs of chewing and crunching, or of stomach acid etching the smaller fish's bones. Then John Long, a Gogo-Formation expert now at Flinders University in Adelaide, examined a fossil placoderm that turned out not to be another fish's dinner

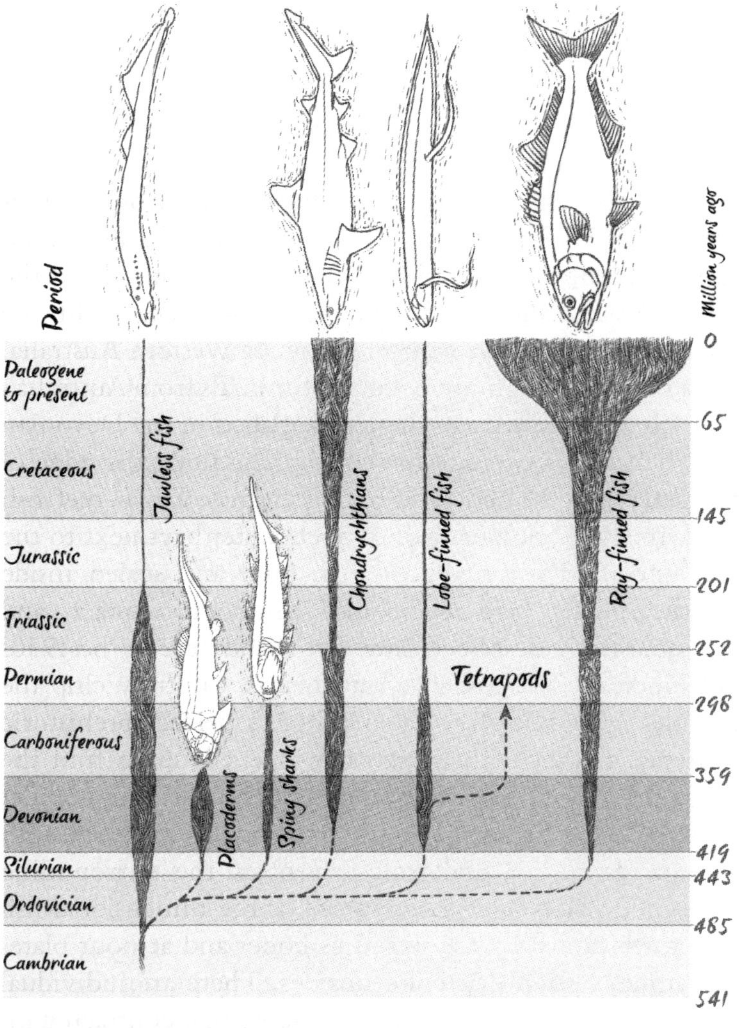

The origins and extinctions of the major fish groups (widths of the spindles roughly indicate the diversity of each group).

but an unborn embryo. His research team found a tiny umbilical cord, with a slight helical twist, connecting the embryo to its mother. In 2008 they named her *Materpiscis attenboroughi* – *Materpiscis* meaning 'mother fish' in Latin, and *attenboroughi* in honour of Sir David Attenborough, who featured the Gogo Formation in his 1970s television series *Life on Earth*.

At a lecture in London in 2010 for the International Commission for Zoological Nomenclature (the naming of species), Attenborough shared his recollections of that filming trip. His Australian collaborators had assured him there would be nothing to see at Gogo because all the good fossils had already been found and carted off to museums. Attenborough insisted that he would nevertheless like to film at the site where these extraordinary fish had been found, and, grudgingly, a helicopter ride was arranged to take his film crew up to the Kimberley.

When they arrived, Attenborough said in his lecture, 'I stepped out of the helicopter and I put my foot on a boulder, and on the boulder was a rectangular scute.' It was undoubtedly the remains of an armoured placoderm, one of the fossils he'd been told were all cleared out. Attenborough described how he turned to his sceptical collaborator and asked what it was. The Australian's response, he said was 'You bastard!' Everyone in the audience roared with laughter at this, and Attenborough chuckled. 'But,' Attenborough added, 'he was decent enough to let me keep it.'

In his lecture, Attenborough went on to describe what happened when, years later, John Long got in touch to tell him about a newly discovered placoderm that would bear his name. Naturally Attenborough was delighted, 'and then I thought, of course, if you have internal fertilisation then you must actually copulate.' He paused for a moment to let this idea sink in. 'This is the first known example of any vertebrate copulating in the history of life, and he named it after me!' Again the audience burst out laughing. 'So,' lamented Attenborough, 'this has been a source of some concern for me.'

David Attenborough has, however, been let off the hook, because a few years after that lecture John Long went on to make a further discovery. On a visit to the University of

Technology in Tallinn, Estonia, he was sifting through a box of placoderm fossils when he spotted an L-shaped bone; he realised it was a sperm-delivering clasper, the same sort of anatomy that male sharks and rays mate with today (although derived from a different part of the body). The discovery sparked a hunt through museum and private fossil collections, which turned out many more male appendages from the same placoderm species. This wasn't the first time people had found placoderm claspers, but these came from the oldest species so far, and probably a few million years older than *Materpiscis attenboroughi*. These, then, are the most ancient genitalia known in the fossil record, and markers of the origin of vertebrate sex. They confirm that internal fertilisation and live birth, as practised by many but not all placoderms (some laid eggs), evolved early on among these fish. Subsequently most living fish, especially teleosts, dispensed with this ancient way of doing things and went back to laying eggs.

Elsewhere, other discoveries have revealed details of the next stage of life for these early fish. In Pennsylvania in 2004, a bypass for the Route 15 highway was cut into the slopes of Pine Hill, revealing fossil-filled rocks that had previously been deep underground. Palaeontologists from the Academy of Natural Sciences in Philadelphia found hundreds of small, hatchling placoderms, with huge eyes and big heads, and deduced that this had been a nursery. Females came to this spot to lay eggs, then left their offspring to it, playing no further role in their upbringing; there were no mature placoderm fossils in the same area. The young succumbed to their habitat drying up; water levels in a quiet backwater rapidly fell and trapped the hatchlings in an isolated pool without enough oxygen to sustain them. Shortly after they died, and before their bodies began to decompose, a layer of mud swept in and settled gently over them, beginning the process of fossilisation and capturing a moment in stone. A similar preserved placoderm nursery has shown up at a quarry in

Belgium. These are the oldest known examples of animals from different generations living in distinct places underwater, just as many species divide up their realms today between young and old.

As well as exquisite fossils like the Gogo fish, Devonian seas were also filled with fish that left very little of themselves behind in the fossil record, but there are other ways of peering into their ancient lives. Thelodonts are jawless Devonian fish that are mostly missing from the fossil record. A few very rare intact fossils reveal they were sometimes spindle-shaped, sometimes flattened with wide mouths, like miniature Whale Sharks, and some had upright bodies with large, lobed tails. On the whole, though, all that remain of the thelodonts are sprinklings of their minute scales.

In 2017, Humberto Ferrón and Héctor Botella from the University of Valencia in Spain took a novel approach to working out details of the thelodonts' lives, despite the scanty fossil evidence. They examined the microscopic shape of their scales and compared them to modern shark denticles, which vary depending on the shark's ecology: where they live, how they move and so on. Making the reasonable assumption that similar associations between scales and habits and habitats applied to thelodonts as in sharks, Ferrón and Botella suggested these ancient fish led a variety of different lifestyles. There were thelodonts that perched on the seabed, hiding in caves and crevices in reefs, as shown by their abrasion-resistant scales; others swam in schools and had spiky scales that deterred ectoparasites from hitching on; and some bore the hallmarks of speed in their drag-resistant scales, etched with ridges and riblets, just as in fast-swimming sharks. There's also a thelodont that had scales similar to those on living bioluminescent sharks, which allow light to pass through the skin. Ferrón and Botella are cautiously waiting for more specimens to study before they go ahead and make any bold claims that thelodonts could glow in the dark.

Thelodonts, as well as galeaspids, osteostracans, conodonts and many other jawless fish, occupy branches down at the base of the fish evolutionary tree, among the direct ancestors of lampreys and hagfish. Exactly how these early branches of the jawless fish should be arranged remains under discussion, which is perhaps no great surprise; after all, this was a very long time ago.* As new fossils are found and new ways of looking at them are invented, additional details are being added to the mix. It's well established that placoderms arose later, snapping their jaws and reaping the benefits of having teeth fixed firmly in place. Then the acanthodians, the 'spiny sharks', arose, some of which were (probably) the direct ancestors of sharks and rays.

All of these fish shared Devonian seas with lungfish, coelacanths, sharks and ray-finned fish. But as this period drew to a close, roughly 360 million years ago, it became clear that this would be the only time in the history of life on Earth when this many deep branches of the fish evolutionary tree would remain intact all at the same time. The reign of all these fish wouldn't last forever.

In the 17th century, when British naturalist John Ray was compiling his unfortunate book *De Historia Piscium* ('The History of Fish') that almost ruined the Royal Society in London, he studied not only living plants and animals but also fossils. And he struggled with questions that would plague him for the rest of his life. How did fossils come to be lodged in rocks, and how were they made?

---

* Imagine time was measured in distance, you set off on a walk into the past and at 100m (330ft) you reached the point when human divers started exploring the seas; to go back 380 million years, you'd have to walk to the moon.

At the time, various theories were proposed to explain the appearance and disappearance of fossils. A widely held idea was that fossils were exuded by the rocks themselves as they tried to impersonate living things. Another was that they were sea creatures that had been swept onto land during the great, biblical flood. John Ray was unconvinced by both these ideas. He was a religious man, and while he conceded that some but not all fossils could have stemmed from the deluge of Noah's time, he'd seen for himself that fossils don't all lie together in the same layer of rocks, as you'd expect if they had all been laid down by a single, catastrophic flood. Instead, he found fossils scattered through discrete beds in many different places he visited. To make matters worse, all that rain and flooding would surely have swept creatures out to sea, not the other way round. He doubted that a great flood would sweep animals from the sea onto land or indeed up mountains.

On his travels, Ray visited the Mediterranean island of Malta, and saw in rocks high in the mountains neat, triangular stones that at the time were known as *glossopetrae* (from Greek words *glossa* and *petra*, meaning 'tongue' and 'stone'). In the middle ages, people believed these tongue-stones held great powers, and wore them as pendants or sewed them into special pockets in their clothes. In the event of a snakebite, you could whip out the triangular stone and press it against the wound in the hope it would save your life. And if you suspected someone was trying to poison you, you could simply drop a *glossopetrae* into your glass of wine as a pre-emptive antidote. Pliny the Elder wrote about these stones falling from the heavens during eclipses of the moon, while others believed that they were the petrified tongues of snakes or dragons. John Ray, on the other hand, could quite clearly see that these stones looked like sharks' teeth.

Earlier in his trip, in Montpellier in France, Ray met with the Danish anatomist Nicolas Steno and it's likely the pair discussed the origins of fossils. Steno is best remembered for his close encounter with a Great White Shark. By the time he saw it, the shark was no more than a dismembered head. The animal had been spotted swimming off the west coast of Italy by fishermen who apparently caught it in a sling and brought to land, tied it to a tree and beat it to death. The shark's head was taken to Florence, where Steno dissected and examined various anatomical details, including the pores dotting its snout, and which were later described by one of Steno's students, Stefano Lorenzini (whom the electro-sensitive ampullae of Lorenzini were named after). Seeing the Great White's teeth up close, Steno was convinced the stony *glossopetrae* must have come from sharks whose teeth fell out a long time ago and were buried in layers of mud that preserved them, changing their chemical composition and turning them to stone.*

Steno wasn't the first to figure out the true identity of tongue-stones, but he made a strong case that began to shift prevailing views on the origins of fossils. Recognising sharks' teeth in rocks was one thing, but John Ray faced a further conundrum in the fossils that look nothing like any known living animals. Could these fossilised animals have lived a long time ago and since become extinct? The idea flew in the face of his religious beliefs. In Ray's view, it was impossible that a benevolent and wise god would let his perfect creations – his creatures – die out. Extinction simply wasn't on the cards. Nevertheless, Ray found a way around the problem and believed that those exotic animals trapped in stone must still exist somewhere on Earth; it was just a matter of time before someone found them. In

* Fossilised shark teeth also helped Nicolas Steno develop a theory for the formation of rocks from sediments piling up, with the oldest layers lower down and newer layers on top. This formed the foundations of stratigraphy, a major branch of geology that's concerned with the layering of rocks.

essence, it's not such an improbable idea. After all, coelacanths were thought to have been extinct for millions of years, right up until a living one was caught off South Africa in 1938. But coelacanths are a highly unusual case. There almost certainly aren't any covert dinosaurs wandering through untouched rainforests, or giant, prehistoric sharks lurking at the bottom of the Marianas Trench. With all the explorers scouring the planet today, someone would probably have spotted one by now.

The idea of extinction only really began to take hold in the following century, thanks to the work of the French zoologist Georges Cuvier. At the same time that he was compiling a catalogue of all the world's living fish in his *Histoires Naturelles des Poissons,* Cuvier was also planning to do the same thing for all the known fossil fish.

In 1831, a few months before Cuvier died, he was visited at the Muséum national d'Histoire naturelle in Paris by a young Swiss scientist called Louis Agassiz. They had been corresponding for a while, and Cuvier was impressed by Agassiz's manuscript on fish from the River Amazon. Agassiz was planning to write a book on the fossil fish of central Europe and was initially worried he might be treading on Cuvier's toes, but after his visit to Paris he came away with a much grander ambition. For several months Cuvier mentored Agassiz and, recognising his skill and devotion to the subject, handed over all his notes and drawings of fossil fish from the Paris museum's fine collections. Agassiz then spent years studying these fossils and many more from across Europe, including from the old red sandstones of Scotland, and between 1833 and 1843 published five lavishly illustrated volumes of *Recherches sur les poissons fossiles* ('Studies of fossil fish'). Across the pages of his books were thousands of detailed drawings of fossils, including some that people thought at the time might be turtles or giant beetles. Agassiz called them placoderms, although he thought they were jawless fish (specimens showing the insides of their skulls and the structure of their jaws were only studied in the 1920s). These strange fish were unlike any still swimming around, and Cuvier would undoubtedly have been convinced they were extinct species.

Years previously, Cuvier had put forward evidence for this theory of extinction. He conducted detailed anatomical studies of elephant bones dug up near Paris. From their size and shape, he concluded they were most definitely not from elephant species still alive in India and Africa. He argued that there couldn't possibly be a third living species hiding somewhere. Elephants are just too big to miss. The bones from Paris must have come from an elephant that was no longer around (he later named them mastodons). Cuvier wrote, 'All of these facts ... seem to me to prove the existence of a world previous to ours, destroyed by some kind of catastrophe.' He came to conclude that repeated revolutions, as he called them, hit many millions of years apart, each time wiping out swathes of life on Earth.

Cuvier's revolutions are now generally known as mass extinctions, and so far in Earth's history there have been five of them. Each episode was triggered in its own way, although often involving rapid changes in global climate, and each dispatched particular groups of living things.

The first mass extinction took place at the end of the Ordovician period, around 443 million years ago, wiping out more than half of all life in the seas. The second event was a series of extinction pulses that swept through the seas towards the end of the Devonian. Immense areas of tropical reefs, like the Gogo Formation, died off. Three-quarters of fish groups went extinct. Jawless fish dwindled. Conodonts suffered major losses; thelodonts were lost; the last of the galeaspids and osteostracans swam around and their giant, armoured heads would be missed from the oceans from then on. Coelacanths became rare, and lungfish were driven out of the sea and only survived in freshwaters. And the placoderms were no more.

The forces behind this colossal upheaval are thought to have come from above the waterline. Around this time, animal life on land was just getting started. Pioneering invertebrates had been crawling around out of water for a while, and amphibians were beginning to pad about on their newly evolved legs. Plants, meanwhile, were exploring the land in a whole new way. They'd been lingering at the waters' edge and in damp places for perhaps 100 million years already, but by the late Devonian they made a break for the open reaches of dry land. For the first time tall trees towered into the sky. Forests spread, and dense canopies of leaves were busy photosynthesising, soaking up carbon dioxide on a vast scale. This thinned the insulating layer of the greenhouse gas – the opposite of what's happening with human carbon emissions today – and the Earth's temperature tumbled into an ice age. Water froze into glaciers, sea levels dropped and shallow seas drained away where so much life had recently thrived.

The greening continents may have turned the oceans green too, ultimately sapping them of oxygen and killing off yet more sea life. Plants' roots pushed deep into rocks, breaking them apart, building up soils and releasing nutrients that washed out to sea. This could have nourished great blooms of planktonic algae, smearing bright, swirling stains across the sea. And when all that algae died and sank to the seabed, bacteria would have decomposed the abundant carcasses, soaking up oxygen from the water and creating dead zones where few living things can survive.*

It's still not fully understood or agreed upon why certain groups of species, and especially among the fish, succumbed to the end-Devonian extinctions that raged for millions of years. What is clear is that the oceans were radically rearranged. The placoderms were no longer the dominant predators, patrolling the open seas and picking prey from

* Similar dead zones are seen in the parts of the ocean today that are overfed with nutrients, chiefly from farm runoff and sewage.

the seabed. In their wake they left great gaps in aquatic ecosystems that were waiting to be filled. And one group of fish in particular made the most of them.

Carnival of the sharks

In the US state of Montana, at a rock formation known as Bear Gulch, are the remains of a shallow bay that once lay at the edge of a vast ocean. For decades, palaeontologists have been chipping away at this 30m- (100ft-) thick limestone deposit that formed in the Carboniferous, around 318 million years ago. The Bear Gulch fossils give an indication of what happened after so much was lost at the end of the Devonian. This was a time when the sharks and their relatives became the leading characters in a curious new aquatic world.

On the bottom of the bay perched *Balanstea*, sharks with stubby tails, flouncy fins and upright bodies shaped like crooked leaves. They were not speedy swimmers but could deftly manoeuvre, grabbing invertebrates encased in hard shells and crunching them between knobbly tooth-plates, fashioned into a beak. In open water, a common sight were large schools of *Falcatus* sharks. It was easy to tell which of them were male and which female. Females, with torpedo-shaped bodies, looked similar to living Spiny Dogfish, although much smaller at only 15cm (6in) long, the size of a hotdog. The males had claspers, modified fins for delivering sperm, and also an additional appendage fashioned from an elongated fin spine fixed to their foreheads and bent forwards as far as the tip of their snout. Richard Lund, a leading specialist in Bear Gulch fossils, has found specimens that hinted at what this spine was used for. A fossilised female *Falcatus* has her jaws clamped tightly to a male's appendage. As fossils, the couple lay together, although the wrong way round for what you might imagine they were up to: the female is on top with her belly pressed against her partner's back. Perhaps this was part of some ancient foreplay.

Another item of bizarre headgear was attached to sharks that were shaped like eels, called *Harpagofututor*. The males had pairs of long antennae sticking up in front of their eyes, forked at the end like crab claws. These, we can presume, were used in a similar way to how living chimaeras use their head-mounted, retractable organs in crucial moments to keep their mates close by.

Swimming around Bear Gulch bay were yet more sharks called *Stethacanthus*. Two species have been found. One was almost 3m (9ft) long, the other was the size of an Atlantic Salmon (70cm, 27in), and both had peculiar things on their heads. Male *Stethacanthus* had what looked like an enormous toothbrush fixed to their dorsal fin, and another brush between their eyes.

Fossils of *Stethacanthus* have been known of for more than a century, but still all we have are wild ideas about why they evolved such bizarre embellishments. Perhaps females picked out the males with the biggest, most impressive head-brushes (male *Stethacanthus* also had long spines, known as fin whips, trailing behind their pectoral fins, which could have played some part in mating rituals). Maybe males postured their bristles at each other while arguing over females, or perhaps they fought in head-to-head battles like deer, only instead of locking antlers they rubbed their brushes together.

The head-brushes probably did have something to do with mating, but there are other suggestions, too. A study published in 1984 points out that the spiny brushes have a similar microstructure to human erectile tissue, so perhaps the *Stethacanthus* brushes were similarly inflatable. Could they have put off predator attacks by engorging their brushes to imitate the giant jaws of a much larger, more dangerous fish? Maybe. Another idea is that these peculiar sharks used their head-brushes to hitch a ride on the bellies of larger sharks, like remoras and shark-suckers do today. If they did, then *Stethacanthus* would have had a unique way of doing it, sticking themselves in place with their very own version of Velcro.

The strange-looking sharks weren't alone in Bear Gulch. Ray-finned fish were there, as they too were expanding their ranges and diversifying after the placoderms vacated the seas. There were various coelacanths and one of the oldest known lampreys, *Hardistiella*, showing that some jawless fish did manage to survive the Devonian extinctions, though they would never again be as widespread as they once were. There were also other, less eccentric sharks. There were chimaeras that looked a lot like the ones alive today, and some species known only from a few scales and teeth, so there's no knowing what these animals looked like or how strange they might have been. And the chondrichthian oddities were by no means confined to Bear Gulch. For millions of years to come, after the Carboniferous period, sharks all across the oceans were still experimenting and coming up with unusual results.

When elegant, spiral-shaped fossils were first found in rocks they were assumed to be ammonites, the extinct relatives of octopuses and nautiluses. They were of a suitable size, often around 20cm (8in) across, and formed an approximate logarithmic spiral (one that expands outwards at a constant rate). Then it was pointed out that they look not so much like seashells as like sharks' teeth. No skeletal remains have come to light, so we can only guess at what the rest of these sharks from the Permian (from around 290 million years ago) looked like. Based on the size of their teeth, they may have been on average around four metres (13ft) long, and perhaps much more than that. An enticing mystery is where on their bodies those jagged spirals belonged.

For more than a century, palaeontologists have imaged various permutations of *Helicoprion,* as the group of extinct species came to be known. The spirals have dangled from the end of a shark's tail or swept backwards on a dorsal fin, or stuck out at the end of an elongated lower jaw like a pizza-cutter. One arrangement even imagines the circular teeth embedded side-on in the flat belly of a stingray. It was only in 2013 that researchers from the Idaho Museum of Natural History published a study of a fossil that shed new light on things. Led by Leif Tapanila, the team took out of storage a fossilised spiral that was originally excavated in the 1950s, and put it inside a CT scanner. Fragments of cartilage still stuck in place, imaged in three dimensions, revealed that *Helicoprion* set their spiralling teeth deep within their lower jaw. It was as if they had a circular saw in their mouths, in roughly the place where your tongue lies. Now imagine your tongue has a Mohican of teeth running down its midline, with new ones being continually made at the back of your throat, nudging older ones ahead as the whole thing spirals downwards and inwards. Inside the *Helicopiron* spiral, the teeth at the centre are the smallest and oldest, made when the animal was young (in a similar way, the central whorl of a mollusc's shell is the oldest part, which it has had since it hatched). With new images and information, Tapanila and the team found clues that *Helicoprion* was more closely related to living chimaeras than to sharks and rays, and what's more, the spirals worked alone. In the upper jaw, it seems *Helicoprion* had no teeth. And so with one mystery solved, another emerged. How exactly did these animals use this singular, toothy whorl? Again, lots of ideas have been offered up and recent studies of a different extinct fish have offered new clues.

Edestus, a close relative of *Helicoprion*, had whorls of teeth in both the upper and lower jaws, although they weren't tight spirals. It had been thought that *Edestus* used their jaws like giant scissors, but this wouldn't have worked because their tooth whorls didn't overlap; imagine using a pair of scissors with the blades bent backwards. Then, in a 2015 study, Wayne Itano from the University of Colorado came up with an alternative idea. His analogy for *Edestus* jaws is a traditional Polynesian weapon called the *leiomano*. This flat wooden paddle, like a giant ping-pong bat, has sharks' teeth fixed to the edge, pointing outwards; it was designed to lacerate the flesh of human enemies. Itano suggested *Edestus* might have used its teeth to slice at soft-bodied prey, like squid. And he thinks it did this by vigorously nodding its head up and down, with its jaws wide open.

These sharp-jawed predators may even have left behind evidence of their feasting. Fossil *Edestus* found in Indiana lie in the same rocks as a mass of mutilated bony fish. There were fish with only heads or only tails; one fish had its tail dangling by a thin strip of skin, and another had a wound that nearly sliced its head clean off. For now there's no firm forensic evidence to blame this massacre on *Edestus*, but they're undoubtedly the chief suspects. And perhaps, in their time, *Helicoprion* did something similar. With their jaws open and their fixed spiral of teeth exposed they could have slashed at schools of squid and fish, as fearsome, head-banging predators.

None of these strange sharks still swim through the oceans today. Most were extinct by the end of the Permian, around 250 million years ago, and they seemed to slip quietly away without unleashing an ecological catastrophe. It would be almost another 200 million years before the oceans were once again shaken up by a mass extinction that changed everything.

Probably the most famous date in the geological calendar is the end of the Cretaceous, around 66 million years ago, when a beloved group of animals bid their final farewell.* As well as the dinosaurs going extinct, this was a mass extinction that also saw off many other animals, including plenty of fish, large ones in particular.

---

* It used to be 65 million years ago, but dating updates tell us we need to add on a million years.

Swimming their final laps of Cretaceous seas were ray-finned fish that looked a lot like tuna and billfish but belonged to a now extinct group, the pachycormids. Most of them were apex predators that raced through open seas; some had long rostra sticking forwards like swordfish or, like sailfish, they had tall dorsal fins, which they may have used to herd schooling fish while they hunted.

Among the pachycormids were filter-feeding giants. *Leedsichthys* was the biggest of them all, and the biggest bony fish ever to evolve, so far as we know. It grew to at least 16m (52ft), slightly longer than a London double-decker bus.* Fossils of these fish were first found in the late 19th century near Peterborough, England, by a farmer, Alfred Nicholson Leeds. Experts originally declared they were the back plates of a stegosaur, but later realised these were in fact the skull bones from a huge fish, which was named *Leedsichthys problematicus*, after the man who found them, and the puzzle of figuring out their true identity.

Until a few years ago, *Leedsichthys* was the only known filter-feeder from the whole of the Mesozoic era (between 252 and 66 million years ago). Splendid as it was, this giant had been consigned to an evolutionary footnote, a brief experiment in filter-feeding that existed for only a few million years. Recently, though, Matt Friedman from Oxford University led a team of palaeontologists that re-examined known fossils and discovered that *Leedsichthys* was not alone. There was in fact a succession of filter-feeding pachycormids that for at least 100 million years

* Previous estimates suggested *Leedsichthys* would have contended with the Blue Whale as the largest animal ever to evolve, at 30m (100ft) long. Recently, though, updated calculations based on incomplete fossil skeletons suggest they probably didn't actually get that big, but at 16m (55ft) *Leedsichthys* would still have been by far the biggest teleost of all time, and the second biggest fish, bigger than your average Basking Shark (15m) and a little smaller than a Whale Shark (20m, 66ft).

were roaming the oceans with their jaws wide open, sifting tiny animals from the water. Their disappearance at the end of the Cretaceous may have created the ecological space for the modern filter-feeding fish to rise to prominence.*

Nevertheless, it wasn't simply the loss of big, conspicuous ray-finned fish like the pachycormids that once again rearranged the living systems of the seas. Other changes took place that paved the way for the rise of modern fish.

Not long ago, an important part of the story of what happened at the end of the Cretaceous was retrieved from the bottom of the sea. Networks of deep-sea drilling programmes have pulled long columns of sediment out of the seabed, far beneath the waves, at points dotted across the globe. These cores drill down hundreds of metres through layers of mud and silt that have accumulated over millions of years. Lodged inside the mud are minute fossils of fish teeth and scales that were scattered across the seabed when the animals died. These microfossils are much more widespread than fossils of whole animals; the odds are extremely slim of a body enduring the ordeals of deep time, of staying intact throughout the fossilisation process and not being crushed and bent out of shape. Teeth and scales are much more durable, and they're also immensely abundant: there can be hundreds of them in a few grams of sediment. From these minute remains, bigger pictures can be drawn of the shifting, changing oceans.

At Scripps Institution of Oceanography in San Diego, Elizabeth Sibert and Richard Norris picked out thousands of microfossils from deep-sea cores. They assembled a timeline of teeth from ray-finned fish and denticles from sharks spanning 45 to 75 million years ago, and they discovered a distinct shift part-way through. In Late Cretaceous sediments more than half the microfossils were

* Manta Rays and Whale Sharks first appeared in the late Palaeocene, roughly 60 million years ago, and Basking Sharks in the mid-Eocene, around 20 million years later.

shark denticles. Then, at the beginning of the next geological period, the Palaeocene, teeth from ray-fins suddenly became more plentiful, outnumbering denticles by two or three times.

This change took place either side of a thin layer of the chemical element iridium, which is rare on Earth but common in meteorites. It's thought this iridium layer was brought in by a gigantic meteorite that struck the Earth 66 million years ago, etching rocks worldwide with an indelible timestamp. Across this iridium boundary, ray-fin teeth not only became more abundant but also three times bigger, going from an average of one to three millimetres (0.04 to 0.1in). This doesn't necessarily mean the fish themselves got bigger – some tiny fish have massive teeth and *vice versa* – but it's a strong indication that fish were adapting to new ways of feeding, and expanding into new habitats.

In their 2015 study, Sibert and Norris go through various possible explanations for this lurch in the ratio of teeth to denticles. Maybe, for some reason, ray-fins simply started growing and shedding more teeth, but that doesn't account for the increase in size. The view Sibert and Norris subscribe to is that 66 million years ago the oceans went through a wholesale transformation.

Before then, ray-fins were relatively rare in the pelagic realm and sharks dominated the gyres that swirled around the Atlantic and Pacific Oceans. For millions of years, this situation seemed stable and unlikely to change. Sediment cores, in particular from the South Pacific, show a flatlining of the ratio of ray-fin teeth to shark denticles, right up to the iridium anomaly; then something jolted the system and the ray-fins quickly took over.

Sibert and Norris point to the mass extinction that brought the Cretaceous period to a close. The meteorite that laid down the iridium layer is at least partly to blame, worsened perhaps by massive volcanic activity that poured carbon dioxide and sulphur dioxide into the atmosphere.

Dark skies loomed and acid seas spread, leaving behind a world vacated of much of its life. And just as in previous mass extinctions, new possibilities unfolded for the few survivors. This time, for some reason that isn't yet fully understood, it wasn't the sharks but the ray-finned fish that muscled their way into the abandoned waterways and flourished.

Previously, it had been well established from the fossil record that most of the major ray-finned fish groups alive today emerged over the last 50 to 100 million years. Some are known from the Cretaceous, like pufferfish, anglerfish and eels, and they were joined later by herring and sardines, minnows and carp, tuna, mackerel, flatfish and many more of the now super-abundant teleosts. The precise timing and the factors involved in the ray-fins' rise to fame had been less clear. Now, this new view from the microfossils narrows things down and hints that the mass extinction played a pivotal role in filling the underwater world with the fish that we see today.

It might have been linked to the disappearance of ammonites and giant ocean-going reptiles, the plesiosaurs and mosasaurs, which were all extinct by the end of the Cretaceous. After they disappeared, ray-finned fish found themselves facing less competition for food and had fewer predators chasing after them.

A parallel story was unfolding at the same time on land. The disappearance of the dinosaurs set the scene for small, furry vertebrates to emerge from their nocturnal hiding places. So, it could be that if a giant meteor hadn't happened to slam into the Earth the way it did 66 million years ago, the oceans, lakes, rivers and streams might not be brimming with so many thousands of fish species that wave their bony rayed fins, and similarly humans might not be here to watch and ponder them.

The Doctor of the sea

Persia, eighth century

The Book of Stones tells the story of a famous natural philosopher who set sail across the ocean to search for a doctor of the sea. He believed that inside the head of this magical fish was a yellow gemstone that could cure all maladies. The fish healed other sea creatures by rubbing its head on their wounds. And the gem also turned silver into gold, which was why the philosopher wanted one for himself.

For weeks the philosopher and his sailors searched and searched before they found a shoal of doctor fish. They cast their nets and caught one, but soon it transformed into a beautiful woman. She spoke a language no one understood and demonstrated her powers, healing the injuries and sickness of the crew. In time, one of the sailors fell in love with her, and she bore him a son who was human in every way except for his shining forehead. All the men thought she was happy living with them on the ship, but one night she escaped and dived into the sea, leaving her son behind.

The sailors continued on their way until they were hit by a dreadful storm. Enormous waves washed over the ship and the philosopher was sure the ship was doomed, until he spotted the strange woman, the Doctor of the sea, floating on the wild waves. The men begged her to help, and she transformed into a giant fish, opening her huge mouth and drinking the sea until the waves were calmed. Her son dived into the sea and followed after her down beneath the waves. When he returned to the ship next day, he too had a gleaming yellow gemstone in his head.

CHAPTER NINE:
FISH SYMPHONIES

CHAPTER NINE

Fish symphonies

On the fringes of the Gulf Stream, off the east coast of Florida, the sea is very deep and very blue. I hold tight to the railing on the fly deck of the dive boat that rolls sharply from side to side, and look down into water that's a thicker, denser colour than I've ever seen. For a moment I imagine that if I leaned over the side and dipped my hand in the water it would come out coated in blue, like paint. Golden fragments of seaweed float by, escapees perhaps from the Sargasso Sea's swirling gyre. I would have been content to stay on deck, watching the sea's colours go by, but there are deeper things for me to see. I pull on my dive gear and jump in. Beneath the waterline, as I kick downwards, the colours lose their intensity and slowly fade away.

Sitting on the sandy seabed at 30m (100ft) is a shipwreck. It's a tanker that was seized in 1989 after US customs found it stuffed with marijuana, and was then deliberately scuttled and sunk to create a new underwater habitat. I aim for the deck that's become fuzzy with a halo of seaweeds, corals and other soft creatures, and hunker down behind the gunwale in a quiet spot away from the current.

Dark shadows lurk nearby, in a hatch in the tanker's superstructure. Before I see the animals inside I hear them, or rather I feel them push pressure pulses into the water that resonate through my body. The bass notes are probably around 50 or 60 hertz, the lower notes on a pipe organ. Another boom and I notice the wreck is vibrating. Then a fish reveals itself, a Goliath Grouper. Its looks as if it was carved from a great chunk of granite; it may well weigh as much as a Grizzly Bear.

Since the wreck was installed on the seabed, the Goliaths have adopted it as a seasonal home where they congregate in the summer months to mate. There are, though, far fewer of these fish across the mid-west Atlantic than there once were. Not so long ago, their meat was canned for dog food, and their carcasses used to smuggle drugs into the US. For decades they've been a favourite of sports fishermen who love to reel them in, hold them up for photos, then throw them back into the sea, already dead. A 2009 study measured the Goliath Grouper's historic decline using archives of trophy photographs taken by sports fishermen; in the 1950s, the catch of Goliaths often outweighed the human passengers on board a sport fishing boat, but their numbers had already been decimated by the late 70s.

Following prohibition on hunting them in US waters in the 1990s, Goliaths seem to be doing a little better, at least in east Florida.* If you venture underwater at the right time of year, there's a good chance you'll find a crowd of these giants and hear the deepest fish voices in all the oceans. It's not clear exactly what meaning lies behind these sonorous calls – a warning, perhaps, or a male showing off to females – but there's little doubt that these big fish are indeed talking to each other.

It would be easy to assume that fish are silent and unhearing creatures. They don't have ears, at least not ones that stick out of their heads. And the sounds of the sea stay trapped below. Most sound waves don't pierce the waterline but bounce back down into the depths. Fish certainly do make and hear sounds, but it's taken a long time for people to realise just how sonic their aquatic world can be, partly

* At the time of writing there are calls to allow fishing of Goliath Groupers once more, despite scientists' warnings that this will once again deplete the spawning population.

because we ourselves are not well adapted to hearing when our ears are full of water. Normally, airborne sound waves travel down a canal towards our inner ears, making our ear drums vibrate, but when that canal is flooded it dampens the quivering membranes, muffling the sound.

The handful of noisy fish that have been known since antiquity are ones that protest loudly when they're taken out of water and dangled in fresh air. Aristotle wrote about fish that call like cuckoos, grunt or make piping sounds; there were also, he said, some sharks that squeak.

Another difficulty in hearing fish sounds underwater is the fact that normally, in air, the slight delay between sound waves reaching our left and right ears tells our brains where the sound is coming from. Travelling so much faster in water, sound waves hit both ears almost simultaneously, making it hard to pinpoint the source. In between my noisy breaths when I'm scuba-diving, there's often a diffuse cloud of sound all around me. It takes something as loud and obvious as a booming Goliath Grouper to give me a good idea of what's going on.

All in all, human ears are not good at picking up and distinguishing the sounds of fish. To make sense of underwater noise, to fully appreciate how talkative fish can be, we need special recording devices to do the listening for us, and it wasn't so long ago that those came along.

In December 1963, a woman with short curly hair sat behind the wheel of a grey Chevrolet sports wagon as she drove north from Rhode Island, along America's eastern seaboard towards Maine. The car was packed with gadgets; there were banks of waterproof microphones, spools of cables hundreds of metres long, two-way radios and walkie-talkies, battery packs and generators, a collapsible aquarium tank made of canvas, and an aluminium boat strapped to the roof. This was a fast-response mobile listening station, on a mission to find noisy fish. The driver's name, it just so happened, was Dr Marie Poland Fish. She was usually known as Bobbie.

As director of a research lab at the University of Rhode Island, Bobbie's work was funded by the US Navy. Back then, the military was keen to know what sounds fish make. Historically, mariners have reported eerie sounds at sea. Moans, thumps and clanking of chains made many think their ships were haunted. This clamour became a major problem in World War II, when the hydrophones of underwater listening stations could no longer detect the distant whir of ship and submarine propellers. Submariners described all sorts of unidentifiable noises: mild beeping and fat-frying, croaking and hammering, whistling and mewing, coal rolling down a metal chute and the tapping of a stick being dragged along a picket fence. At times the racket even drowned out the biggest battleships, disabling an important part of wartime surveillance.

Following some initial investigations, it became clear that some of the noise came down to waves, wind and tides, but animals were chiefly to blame. Fish were so noisy they triggered underwater bombs, which were only supposed to detonate at the sounds and vibrations of a nearby submarine. There was obvious strategic advantage to be gained from knowing more about the hubbub of sea life, including when and where it was noisiest, which is where Bobbie Fish came in.

When the war finished, and for the next 20 years, she set out to record and identify these unseen sound-makers, most of them fish. Using hydrophones developed as part of the war effort, she fixed long-term listening stations in rivers and bays to gather ambient sounds of the underwater world. Between 1959 and 1967, a research boat went out every week into Narragansett Bay, off the coast of Rhode Island, and brought back fish to Bobbie's lab, where she recorded their voices. With hydrophones dangling in the tank, captive fish were recorded at different times of day and in different circumstances: when they were new to the tank, when other fish were added and things got more crowded and boisterous. For the fish that stayed stubbornly quiet Bobbie administered a mild electric shock, which often elicited a prolonged audible response. Experts have criticised this approach, because it's possible some of those sounds may not have been natural noises the fish would make without being zapped.

The team also auditioned fish in other research labs and aquaria in America and the Caribbean, and they hit the road in Bobbie's customised car. On that December trip, the car's first excursion, Bobbie was heading for Boothbay Harbour in Maine to record a chorus of winter fish. With her were oceanographer Paul Perkins and electrical engineer William Mowbray, whose voices you can hear on the archived recordings, announcing the name of each fish.

In 1970 she co-wrote with Mowbray *Sounds of Western North Atlantic Fishes,* a book filled with spectrograms, showing the shape and texture of fish sounds.* The mix of tones and pitches in a fish's voice were plotted in charts, revealing the intricate differences between croaks and barks, hums and grunts. The book contains spectrograms from the fish Bobbie recorded in Boothbay Harbour, like the Pollock that was lowered into the canvas tank and made thumping sounds when it was handled; its spectrogram shows repeated smears of sound, like a comb dragged through paint. Another Boothbay fish from 1963 was the Grubby, a type of sculpin, whose spectrogram has two clean lines, one lower- and one higher-pitched, both lasting for four seconds, then again for two. The book also features the voice of an Ocean Sunfish that was found just outside Narragansett Bay and

---

* Spectrograms plot sound wave frequency on the *y* axis against time on the *x* axis; the higher the frequency of the sound wave, the higher the pitch.

held in a sea pen; it made rasping grunts like a pig, which became louder and more frequent the more it was handled. A Goliath Grouper in Puerto Rico let off a tremendous boom whenever it was prodded, producing a spectrogram that looks like a series of short strokes of a soft paintbrush; another in the Bahamas stayed quiet, although it did, on one occasion, almost swallow the hydrophone in its enormous mouth.

These findings helped navy personnel to tune out the sounds of fish and once more tune into the sounds of their enemies. Bobbie Fish had used her listening devices and analytical tools to pick out individual voices from the underwater cacophony. She had shown that it's not just a few fish species that are noisy, but hundreds of them. And as she wrote in the introduction to *Sounds of Western North Atlantic Fishes*, 'The mechanisms for sound production in fish are varied and often ingenious.'

Just like their abilities to make electricity and toxins and light, fish have evolved ways of making sounds many times. They've transformed different parts of their bodies into noise-making instruments. Fish gnash their teeth to make rasping sounds; anemone fish have tendons that snap their jaws shut and their chattering teeth make chirps and pops; coral reef-dwellers called grunts get their name from the grunting sounds they make by grinding their second set of teeth together, at the back of their throats (their pharyngeal teeth); porcupinefish rub their toothless jaw bones together, making a sound like a rusty hinge. Seahorses click when they flick their head upwards to catch plankton and two bones at the back of their skull push past each other; they also purr and growl, sounds generated somehow in their cheeks (with a mechanism that isn't yet fully understood). Sculpins use muscles to rattle their pectoral girdle.* Channel Catfish, a common sight and sound in North American rivers and lakes, rub serrated fin-spines over a roughened patch on another bone, in a similar way to how crickets and grasshoppers sing and chirp. Croaking Gouramis are popular pets and native to still waters of Southeast Asia, in ponds and paddy fields; their croaking name comes from the sounds they make when they beat their pectoral fins against specialised tendons, like strumming a guitar.

The most common body part that fish use to make sounds is their swim bladder. This internal gas balloon – commonly sausage-shaped or like a modelling balloon that's been twisted part-way down into two lobes – first evolved in fish as a lung to breathe through, then was co-opted as a flotation device, and later adapted in many ways to emit sounds.

Think of all the different sounds you can make with a regular balloon. You can drum your fingers against it, make it squeak by rubbing its surface against something else, or you can let a trickle of air out in a squeaky whine; fish do all these things and many more. The only thing they don't do (on purpose) is pop their swim bladders to make a loud bang.

Many fish have sonic muscles that vibrate against the swim bladder, making it buzz and hum as it contracts and expands. Some have muscles that stretch the swim bladder forwards then let it go, so it pings back into place. Triggerfish have a drum on each side of their body where the swim bladder pushes up against a patch of large scales called scutes. Struck by the pectoral fins, the scutes bend inwards, then pop back into shape, making a drumming sound. Toadfish, some of the noisiest of all the fish, call out like a foghorn by rapidly vibrating their heart-shaped swim bladders. Two sets of sonic muscles make each portion of the toadfish's bladder buzz at different rates, producing complex sounds

* The pectoral girdle in humans is made up of the shoulder blade and collar bone.

similar to the wailing cries of a human baby, sounds that are hard to ignore (especially if you are a female Toadfish).*

Learning precisely why fish make sounds and what uses they serve involves not just listening to fish but also watching them. Film footage reveals that many species use sounds as alarm calls, shouting aggressively when picking fights and screaming to startle predators. Amazonian Red Bellied Piranhas yell at each other with a certain level of sophistication, using three distinct calls in different circumstances. During head-to-head encounters before a fight breaks out, the fish emit sharp repetitive barks as a warning of what lies in store if their opponent doesn't back off; it's a piranha version of trash-talk. Then, when a skirmish breaks out, especially competing over food, the piranhas make a deeper thudding sound while they circle aggressively and take bites out of one another. Both of these hostile noises are made by muscles vibrating the swim bladder. A third call is a volley of much higher-pitched sounds, made by gnashing their teeth. When a victorious piranha chases off another, it makes this sound, presumably to say, 'I win. You lose. Don't come back.'

Pacific and Atlantic Herring are far less furious fish. They seem to communicate with gentle streams of bubbles trickling from their swim bladder and out through their anus. Emanating from the trails of bubbles come pulses of sound, up to seven seconds long, which researchers named Fast Repetitive Ticks or FRTs for short. Film footage shot in huge aquarium tanks in the dark with infrared cameras shows young herring swimming around in loose shoals, making bubbles. With a screen across the top, blocking their access to the air, the herring quieten down after a few nights, probably because they can't refill their swim bladders

---

* These are known as non-linear sounds. Filmmakers make use of the way these sounds heighten an audience's emotional response by adding them to sound tracks at key moments; Alfred Hitchcock used them in the shower scene in *Psycho*.

by gulping air from the surface into their stomachs, and they run out of farts.* One idea is that they use bubbles to maintain contact with their shoal-mates at night. When the lights come on and they can see each other again, the herring fall silent. They're the only animals known to communicate with flatulence.

Often, the rowdiest periods underwater are when fish are courting and mating. Deep beneath the springtime waves of the North Atlantic, male Haddock swim close to the seabed in tight circles and figures of eight, and they emit slow repeated knocks (they will also perform their mating rituals in large aquarium tanks, which is how we know they do this). The noises are important because it's dark and murky down there. Females hear the calling males and come to investigate. A male will then trail after a female. He'll also swim ahead and get in her way, flicking his fins at her and showing the three spots he's drawn across his flank, by shifting pigments around inside the chromatophore cells in his skin; normally there's just one dark mark, his 'Devil's Thumbprint', as it was traditionally called. All the while he speeds up the pace of his knocking. He begins to sound like the roaring engine of a motorbike as the knocks blend together into a constant, loud hum. For at least ten or twenty minutes he can hum non-stop. Eventually, if he's lucky, the female swims upwards and they press their bodies together into a tight clinch. The male's voice reaches a quavering finale as he releases a cloud of sperm and, perhaps signalled by his climactic cry, the female puts thousands of eggs into the water. Then the male falls silent, and the pair break apart and swim off.

* Herring and other so-called physostomous fish (including birchirs, gars, some catfish, eels and trout) have a pneumatic duct between the gut and the swim bladder, allowing them to fill up, fart or burp out gas from the bladder. Other fish, the physoclisti, have lost that connection and instead have a gas gland that pumps gas more slowly in and out of the swim bladder.

This is what Haddock are usually up to when they're scooped from the sea by North Atlantic trawlers that target their spawning sites.

Humans also have their own mating rituals involving fish swim bladders. A century or so ago in Europe, they were fashioned into reusable condoms. Catfish and sturgeon swim bladders were, apparently, a popular size and they needed to be tied in place with a ribbon. In China, soup made from dried swim bladder is considered an arousing delicacy. One species in particular, giant croakers called Totoaba, live only in the Sea of Cortez, and have a sky-high price ticket on their swim bladders (or maws). The illegal trade is taking so many fish they're now critically endangered, as is the world's smallest dolphin, the Vaquita, another endemic to this small sea in Mexico. Vaquita get tangled and drowned in the gill nets set to catch the valuable Totoaba. It might not be long before both species go extinct, and all because of soup.* There are people who pay a lot of money to slurp a bowl laced with Totoaba maw, believing it boosts fertility and contains potent aphrodisiac powers.

Fish swim bladders are still used today to clarify beer. Rich in the protein collagen, dried swim bladder speeds up the rate that yeast clumps together, so it settles out of beer quickly, leaving it sparkly and clear. British breweries originally used sturgeon swim bladders imported from Russia, a by-product of the caviar industry. With rising prices an alternative was sought and in 1795 Scottish inventor William Murdoch showed that much cheaper swim bladders from Atlantic Cod work just as well. By the 19th century, pale ales became popular instead of dark

* At the time of writing, there are estimated to be fewer than 30 Vaquita alive, and efforts to breed them in captivity have recently failed.

porters and stouts, as pub-goers quaffed transparent pints in see-through glasses rather than china and metal tankards. Lately, there's been a backlash against using swim bladders, also known as isinglass, and more breweries are using vegan-friendly alternatives like seaweed, or they're learning to be patient and wait longer for yeast to settle by itself (or they're simply persuading customers to accept cloudy pints).

A whiff of magic surrounds another piece of the fish's sonic equipment. For a long time, people have been rather obsessed with the mystical and curative powers of stones allegedly found inside animals, including snakestones and toadstones,* and fish are no exception. Lodged deep inside fish heads are small, hard stones known as otoliths (from Greek words *oto* and *lithos*, meaning 'ear' and 'stone'). In the first century, Pliny the Elder wrote about these stones being used as charms against swellings in the groin and pains in the eye. Sixteenth- and seventeenth-century literature refers to them being ground up, mixed in wine and used to treat kidney stones or nose bleeds; you can wear one as an amulet to ward off malaria. And naturally, as with so many animal potions, it was said that fish stones could enhance the libido. In his 1502 book *Mirror of Stones*, Italian astronomer and mineralogist Camillus Leonardus wrote of magicians who professed that otoliths 'excited Luxury in the Day' – in other words, they can lead to amorous behaviour, for some reason before the sun sets.

Today, otoliths continue to be widely treasured and put to many uses. Fishing communities in Iceland, Brazil and Turkey still use ground otoliths in folk remedies to treat urinary infections and asthma. Spanish fishermen keep otoliths in their pockets to protect them from storms at sea. And in North America, beach combers on the shores of Lake Erie may come across otoliths that came from a

* Which were nothing to do with toads but were in fact the fossilised teeth of extinct *Lepidotes* fish.

Sheephead, a species of croaker. These fish have pairs of otoliths that form a mirror image of each other, one from each side of their head. Some have a J-shaped groove running across them and, supposedly, bringing joy. Otoliths impressed with the letter L are known as lucky stones and will bring good luck, or perhaps even love.

There's no evidence that otoliths hold any real healing or lucky power for humans, but for fish they play an important role in hearing. Because fish swim in water that's a similar density to their bodies, sound waves tend to pass straight through them. To make up for this they have large granules inside their inner ears made of calcium carbonate, the same material that seashells are made of. Otoliths are denser than water and the rest of the fish's body, and they move more slowly in response to a sound wave, rather like the white plastic flakes inside a snow globe when you shake it. The structure of a fish's inner ear is similar to our own. They have fluid-filled chambers, similar to the human cochlea, lined with sensory hairs. Inside each chamber is an otolith.* As the otoliths vibrate against the sensory hairs they trigger nerve signals to the brain. Also, as the otoliths sink down under gravity, they tell the fish which way is up; flying fish have especially big otoliths, perhaps because their sense of balance is so important while they glide through the air.

If all you have is a fish's ear-stone, there's still a lot you can know about the animal it came from. Similar to molluscs making their external shells, fish continually lay down new layers of calcium carbonate inside their ears. With the aid of a microscope, you can count the layers and know how old a fish was when it died. With a lot of patience you can distinguish the protein matrix sandwiched between daily layers of calcium carbonate, and work out how many days the fish lived for. You can identify a fish species from

* Ray-finned fish have six otoliths; lampreys have four, hagfish have two. Sharks' otoliths are the size of sand grains.

its ear-stones: some look like sea-smoothed chips of beach glass, and others like jagged rice puffs. The outer edge of an otolith can look a little like waves on the surface of the sea, which might explain another long-standing belief that a fish's head-stones can forecast conditions at sea. In reality, otoliths can't predict the future, but they can tell us about the past.

There's a chemical story written into every fish's ear-stone that records details of its life. As they grow, otoliths pick up minute traces of other elements from the water, like barium and magnesium, and the amounts vary from place to place in oceans and rivers. By measuring these chemicals locked up in otoliths, it's possible to work out what the fish ate and where it swam at different times of its life (like the giant Amazonian catfish), even the temperature of the water. And the fish doesn't need to have died recently. Otoliths are so dense and tough they commonly hang around long after the rest of the fish has rotted away, and they fossilise well. Palaeontologists have extracted the stories of fish from otoliths that fossilised hundreds of millions of years ago, and used them to work out the temperature of ancient oceans.

As well as their ears, fish have an entirely separate series of sensory organs laced across their heads and flanks. Known as the lateral line, this effectively converts their whole body into a giant ear. It's an ancient structure that evolved very early on; the oldest known fossilised lateral lines are in jawless fish from the Ordovician period (around 480 million years ago) and all living fish groups have them.*

The basic units of the system are structures known as neuromasts, which that contain tiny hairs that fire nerves when they bend, working in essentially in the same way as

* The only other animals that have an equivalent system are amphibians, and mostly just in the larvae.

hairs in the inner ear. Neuromasts can either sit on the skin surface or lie inside tubes that run under a fish's skin and scales. Lines of dots along a fish's flank shows the points where water enters these tubes. The system lets fish sense the flow of water over their body and detect vibrations close by, at one or two body lengths away. Hunting fish pinpoint the noisy vibrations of insects that fall into the water and kick and struggle on the surface. They also track ghostly imprints that other animals leave behind as they push their way through the water. After a creature passes by, it sheds a wake that lingers for some time. And by the quivering of its lateral line, a fish can know the size and speed, and even the particular swimming stroke of that animal, and decide whether to follow in pursuit or make a quick escape.

Lateral lines are especially important for fish that live in the dark and have no eyes, such as Mexican Tetras. Like other, seeing fish, they use their lateral lines to locate moving objects that swish past. Blind fish can also detect stationary objects, like the walls of their cave, by sucking water into their mouths and sensing any pressure disturbances in the flow that tells them when they're about to bump into something. The closer they get to an object, the faster they open and close their mouth, presumably to create a faster flow and gain more information about what's out there. Bats flying through these underground caves use ultrasonic beams to hunt and navigate, while down below fish interrogate the cave's waterways with their equivalent of echolocation.

The fish that hear the best are the ones that don't rely solely on otoliths or the lateral line, but have a helping hand from their swim bladders. These compressible balloons of gas not only make sounds but detect them too, by vibrating when the pressure waves of sounds pass through. It's another big secret behind the immense success of fish: thousands of species have modified their swim bladders into listening devices.

One way they do this involves a chain of jiggling bones, adapted from four or five vertebrae of the neck, that connects the swim bladder to the inner ear.* These important little bones boost the hearing of one in four fish species, and they're thought to be a key feature underpinning the immense success of the fish that dominate freshwater ecosystems – the minnows, catfish, carp, loaches, knifefish and piranhas.† Inland waters are often murky and difficult to see through, so sound and hearing play a crucial role in the lives of many freshwater fish species.

Other fish have extensions of their swim bladders that connect to their lateral line, or poke through the skull and connect directly to the inner ear. Herring, menhaden, sardines, shad and anchovies all hear this way, and among them are the champions in high-pitched hearing.

Blueback Herring and American Shad swimming in silvery schools off North America's eastern seaboard, and Gulf Menhaden in the Gulf of Mexico, can all hear higher-pitched sounds than any other fish. Captive specimens have been trained with weak electric shocks to lower their heart rate when they hear sounds. This revealed they can detect frequencies up to 180kHz (an average human hears between 20Hz and 20kHz), making them one of only a handful of animals that are known to hear ultrasound, along with bats and cetaceans.

A less invasive fish-hearing test involves placing surface electrodes on their skin to record the firing of auditory nerves when different sounds are played through underwater speakers (this Auditory Brainstem Response, or ABR, is used to test hearing in young human children, although usually not when they're underwater). Wired up

---

* These bones, known as the Weberian apparatus, do a similar thing in as the incus, malleus and stapes in our own ears, which transmit sound waves from the eardrum to the inner ear.

† These and many other fish belong to a group known as the otophysans, which account for more than 60 per cent of all freshwater fish species.

in this way, Goldfish have shown they can hear up to 4kHz, the highest key on a standard piano, as can most fish with swim-bladder-to-ear connections; unconnected fish can only hear up to around 1kHz.

All these tests, on herring, shad, Goldfish and many other fish, pose the same question: what are these fish listening to?

The herring and shad don't make ultra high-pitched sounds, so at these higher registers they can't be listening to each other. Their discerning hearing may in fact have evolved because it lets them eavesdrop on dolphins when they use beams of ultrasound to find prey, including these schooling fish. There are moths that do the same, listening out for bat sonar to avoid getting eaten. Studies suggest that herring and shad can hear dolphins from at least 100m (330ft) away. Likewise, Goldfish aren't listening to each other because they themselves, as far as we know, are silent. They too may be listening out for the sounds of predators, or they could be listening to the assorted soundscapes of their underwater world.

When Bobbie Fish set about recording and analysing the sounds of fish, her main aim was to tease apart the cacophony of underwater noise, to identify voices and assign them to certain species. Since then biologists have, for the most part, continued to focus on the sounds that individual fish make and hear. Gradually, though, a new approach is emerging, as more people are beginning to listen to the entire aquatic symphony.

The world is bathed in light from the sun, and it's also bathed in sound. Underwater, this soundscape may at first seem like a disorderly din, but there's more to it than that. Off the coast of Western Australia, a series of waterproof microphones have recorded distinct dawn and dusk choruses, lasting for hours at a time. These were the sounds of thousands of fish, calling to each other, fighting, flirting, mating and eating at those most active times of day. There is structure in this noisy world.

In the cool, fish-rich rocky reefs off New Zealand's North Island, another set of listening devices revealed that different habitats have their own particular sounds and a unique acoustic signature. By listening, it's possible to tell apart a rocky reef covered in seaweed from one inhabited by sea urchins; as they graze and scrape the rocks with their teeth, the urchins' shells resonate like bells.

Much remains unknown about how fish listen to these ambient sounds. It could be that they try to tune it out so they can hear each other, like having a conversation at a loud party. But there are clues that the backdrop of noise matters to them, that fish listen in and extract useful information from the sonic miscellany.

Nocturnal sounds may be especially important. In shallow tropical seas, many fish are on the move between day and night. During the day, some hide and rest in patches of coral reef or among mangrove tree roots, then as night falls they swim to nearby seagrass meadows to feed. Most make their move when it's dark in the hope they'll go unseen by the most dangerous predators, the bigger fish that hunt by sight. Similarly, newborn fish spend their first days and weeks in open water, again to avoid the reef's many hungry mouths. In time, the young ones' muscles and fins become strong enough to push against tides and currents. Only then do they turn around and begin a long swim back home, guided at night by an inbuilt magnetic compass and during the day by a celestial compass, pinpointing the position of the tropical sun beaming down on the water. As they get closer, the young fish zero in on their native habitat, following their noses and also their ears, listening for the sounds that could act as beacons guiding the travelling fish through the dark.

To investigate this idea, Craig Radford from the University of Auckland in New Zealand led a research team who built small, identical piles of coral rubble, spaced out across shallow waters around Lizard Island on Australia's Great Barrier Reef. Through underwater speakers suspended over each rubble pile they played back soundtracks recorded in different habitats. The morning after a noisy night, Radford and his team counted up the young fish that had arrived on each rubble pile and found that some did indeed seem to be lured by the sounds of certain habitats. Young damselfish headed for rubble piles that sounded like a fringing reef (dominated by the popping and cracking of pistol shrimp as they snapped their claws) and young bream were drawn to the piles that sounded like an open lagoon; far fewer fish were enticed by the sound of silence, played back to them in the control rubble piles. It's still early days, but it seems likely that fish can distinguish between the sounds of different places underwater, and follow their ears to the spot they most want to be.

These habitat soundscapes are subtly composed. Recent studies are revealing that far from this being an impromptu free-for-all, fish don't simply yell and shout however and whenever they want: they fit their voices together like an orchestra of instruments in a melodic musical score.

One such study took place off the KwaZulu Natal coast of South Africa, in the Indian Ocean, a short way south of the Mozambique border. Just offshore, steep canyons carve into the seabed. A hundred metres (330ft) down, in a cave where coelacanths live, a team of European researchers led by Laëtitia Ruppé wedged a small recording device into a crevice in the wall. After two months, the team fetched the device and listened to the sounds of the cave-dwellers. Previously, South African biologists inside mini-submarines had visited caves in the area and seen hundreds of fish species living down there, including sound-making groupers, soldierfish and toadfish.* So it was perhaps no surprise when the cave recordings played back thousands of

---

* It's not clear whether coelacanths are vocal or stay quiet among their chattering neighbours.

noises, many of them fish voices. What was surprising was the patterns those voices made.

Taking the most obvious voices and plotting them on spectrograms like the ones in Bobbie Fish's book, Ruppé's team found that, at night, fish were acoustically avoiding each other. In two dimensions, of pitch and time, each voice occupied its own space on the spectrograph, like pieces of a sonic jigsaw; different fish called at different times or different pitches, building up distinct layers of sound. There were deep, isolated booms, low and long tones and clear, coarse pulses, pops, grunts and high-pitched whistles. The species awake during the day produced more jumbled sounds, perhaps because they could see each other and combine their calls with gestures; when they call, they can swim and flick their fins in eye-catching ways, like shouting to a friend on the other side of a busy room and waving at the same time to catch their attention. In the dark of night, when fish can't see each other, it matters more if they have overlapping, clashing calls. Nocturnal species make sure their voices don't drown each other out.

These fish are partitioning sound in the same way they divide up many other aspects of their ecosystem. Within a community, species evolve to eat different foods and they split up the physical space they occupy; now it's becoming clear that species also set out and establish their own vocal territories.

The ecology of sound is still a relatively new idea, and so far has mostly been applied to terrestrial ecosystems. There are various birds, insects and frogs that similarly divide up their soundscapes and avoid masking each other's calls. Studies on land also point to the problems that unfold for these vocal species when the world becomes noisier with human sounds. Traffic makes it difficult for birds to hear each other and they can miss important messages, particularly during mating times. It's too early to say whether fish will suffer as we fill up the oceans with our human sounds, from shipping traffic, seismic surveys, underwater sonar and thousands of offshore oil and gas platforms. Marine mammals are the focus of most investigations into underwater noise pollution. Fish studies are few and far between. But the chances are there are many fish out there whose lives are shaped by sound, fish that are doing their best to talk and make themselves heard in the clamour of an increasingly noisy world.

The fish and the golden shoe

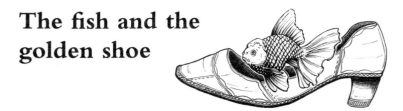

China, T'ang dynasty, ninth century

A girl called Sheh Hsien lived in a little house in the mountains with her cruel stepmother and stepsisters. They would send Sheh Hsien to gather wood from dangerous, faraway forests and to carry water from the deepest wells. One day, she pulled up her bucket and saw inside a small, shining Goldfish. It had red fins and golden eyes, and was as long as her finger. She took the Goldfish home and put it in a basin of water, where it swam round and round in circles. She fed it scraps of food and every day the Goldfish grew bigger, so she had to find larger and larger basins to keep it in. Eventually she didn't have a vessel big enough, so she took the Goldfish outside and let it go in the pond behind her house. Whenever she came to feed it, the Goldfish would swim over and rest its head on the side of the pond. But it only came to Sheh Hsien, and no one else.

When the stepmother saw this big, fine Goldfish she was jealous and wanted it for herself. So she sent Sheh Hsien to fetch water from a well far away, then dressed in her stepdaughter's cloak and went to the pond. The Goldfish came to the stepmother, and straight away she killed it with a sharp knife. She cooked the enormous fish over a great fire. 'That was the most delicious fish I have ever eaten,' she thought to herself, as she buried the Goldfish's bones in a pile of dung.

When Sheh Hsien came back from the well and saw her Goldfish was gone, she sat down and wept. Just then, an old man wandered past who reminded her of her dead father. 'The fish's bones are in that dung pile,' the old man said. 'Go and find them and place them under your pillow.

Then, if you ever want for anything at all, ask the fish and your wish will come true.'

The old man bade her farewell, and Sheh Hsien did as he said. With the bones under her pillow she began to ask the Goldfish for many things and soon she had beautiful clothes and shoes, jewellery and pearls, and the finest food to eat. She missed the Goldfish, and hated her stepmother more for than ever for stealing and eating it, but she was thankful for everything it was now bringing her.

A while later it was the Festival of the Mountain, but the stepmother forbade Sheh Hsien from going. When everyone had left the house, Sheh Hsien dressed herself in a blue silk dress and golden shoes, and sneaked out to join the party. She hadn't been there long when one of her stepsisters spied her and Sheh Hsien ran away. As she fled she tripped and one of her golden shoes fell off.

A crow picked up the glittering shoe, then flew off across a great sea and dropped it on an island that belonged to a powerful king. When the king saw the delicate shoe he ordered his men to find its owner. All the island's women tried on the shoe, but their feet were too big. The king sent his men to search houses far and wide, until they came to Sheh Hsien's house on the mountain and found a single, matching golden shoe. She slipped her feet into both tiny shoes and the king fell instantly in love with her. After that, the king and Sheh Hsien were married and lived happily together on his island, and Sheh Hsien always remembered her lucky Goldfish.

CHAPTER TEN:
(RE)THINKING FISH

CHAPTER TEN

(Re)thinking fish

A small male wrasse, a restless streak of blue, white and black, is coming to the end of another busy day on the reef, and still a few more fish are waiting for his services. A rabbitfish hovers motionless at the front of a short queue of five or six fish. She has all her fins splayed out and her bucktoothed mouth open as if she's suffering from a nasty shock. In fact she's quite relaxed. The wrasse and the rabbitfish know each other well. Today alone they've already met at least a hundred times, the larger rabbitfish returning again and again, infested with more parasites.

The clinging bloodsuckers are minute crustaceans called gnathiid isopods, which dart out of the reef and fix themselves to passing fish. Like aquatic ticks they will suck a fish's blood for about an hour before dropping off. Rather than surrendering their bodily fluids, fish prefer to have these freeloaders unhitched by wrasse, who have become the chief cleaners on the reef. And it's a task that requires a surprising amount of brainpower.

Hundreds of client fish from dozens of species make regular visits to the cleaning station. The cleaner wrasse has memorised them all, and tailors his services to each one. His livelihood depends on his refined social skills, of cooperation and communication and even his cunning ability to manipulate other fish. A staple diet of parasites, thousands every day, keeps him well fed, but in truth these aren't the wrasse's favourite food. He much prefers a nutritious bite of fish skin or the gummy mucus that coats a fish's body.

One thing he's after is sunscreen to help block the sun's harmful ultraviolet rays, which flood shallow, tropical seas.

Fish can't make sunscreen themselves. Most obtain it from microbes in their food. The molecules pass through the fish's digestive system and are secreted in a protective layer of mucus on the skin; fish drink sunscreen instead of rubbing it directly on their skin, as sunbathing humans do. Another way to get sunscreen is to lick and slurp it off another fish, but the cleaner wrasse knows there are only certain situations when he can get away with that. To maintain his territory on the crowded reef and win the trust of other fish, the cleaner must do a good job of removing parasites, and can't push his luck. An injured client will swim off and may never come back. And if other fish waiting in line catch sight of the cleaner wrasse cheating and eating mucus instead of parasites, they may well leave and find another cleaning station.

As the wrasse surveys the rabbitfish it's getting late, and only a small audience is queuing to be cleaned before nightfall. There's a dusky brown damselfish on her third visit of the day. The wrasse knows he doesn't have to behave himself in front of this client because she never strays far from her small farm of algae, and there are no other cleaning stations within easy reach. Another waiting client is a surgeonfish, a harmless herbivore he hasn't seen before, and the wrasse decides to take a chance. He makes a brief show of searching once again across the blue and yellow honeycomb of the rabbitfish's body, picks off two parasites, then goes for it. The wrasse takes a bite of skin and mucus. The rabbitfish flinches, feeling the scratch of teeth. But just as quickly, the wrasse rubs his fins over her back and belly. It's an apology of sorts, soothing and calming her until the disgruntled rabbitfish seems to drift into a blissful stupor. Levels of stress hormones in her bloodstream dip a little. It's probably one reason why she keeps coming back to this cleaning station, even though she knows the wrasse does occasionally cheat.

The mood across the reef then shifts as a new client arrives, a large grouper. Instantly, the wrasse knows this fish is important: she's big and probably has a good collection of isopods to pick off. More importantly, though, she's a predator and could quite easily choose to snack on the little fish during the cleaning ritual. Somehow the wrasse senses the grouper hasn't eaten in a while, and takes extra special care.

It's time to dance.

The wrasse beats his tail from side to side, then shimmies his fins across the predator's stocky body, more than ten times his size. The grouper's gaping mouth yawns open and the wrasse swims right in. He diligently pecks around sharp teeth that are fearful weapons, perfect for impaling small, delicious fish. For now peace is maintained, probably because the wrasse continues to stroke and massage the grouper, pacifying her so that for a while she has no interest in hunting. The two fish, predator and cleaner, strike a deal, and both stick to their side of the bargain. The social bonds between them are strong but not unbreakable. If they met in any other circumstances things would be very different. But here at the cleaning station the wrasse is granted immunity, so long as he restrains himself and eats only parasites.

All the parts of this scene, and many others like it, have played out in front of biologists who have spent countless hours swimming over wild reefs and peering through glass aquarium walls, watching fish interact with one another. These scientists have counted cleaner wrasse picking off hundreds and thousands of parasites; they've devised experiments to test how cleaners and clients decide what to do, and to grasp how the fish recognise each other; they've watched fish dance and cheat and, yes, apologise. Studies like these aren't only showing how fish on coral reefs cooperate to stay clean and healthy, they're also revealing details of complex, clever lives of fish that have long remained overlooked.

A time-worn portrayal of fish sees them as simple-minded creatures, governed by innate reflexes and incapable of thinking for themselves. This perspective is driven by human-centric studies that search among our closest, mammalian relatives for clues as to how and why our own brains evolved. But it's a blinkered outlook that's pulled attention away from animals deemed too distant and different from us to be brainy. By looking at fish in new ways and asking the right questions, biologists are increasingly realising that fish think and solve problems with surprising sophistication. Entrenched ideas about water-dwelling vertebrates are shifting, and at the same time fish are bringing into focus a broader, richer view of what it means to be intelligent.

Most animals have some basic level of cognition; they can sense the world around them in various ways, they gather information, process and store it. More advanced cognition – intelligence, if you like – requires learning from the past and using stored knowledge to solve new problems in the future. Instead of hard-wired instructions on how to live, these are steps towards a more flexible, adaptable approach to coping with a changing world.

Intelligence has always been a slippery concept to define, but if we drew up a list of important signs of intelligent life, fish would tick many of them off. Cleaner wrasse, as we've seen, communicate with members of the same and other species, and they have excellent long-term memories. They manipulate their clients by massaging them, and have a sense of the other fish's motivations. Will this client bolt and never come back? Or does it have no option other than to stick around?

The cleaners also manipulate each other. Male and female wrasse often have overlapping territories, and frequently offer their services together. Operating alone, a female will indulge now and then in some mucus-nibbling, but she soon learns not to when a male is around. Whenever she cheats and sends a client off in a huff, the outraged male cleaner delivers a punishment, aggressively chasing and nipping at her. All he gains from her cheating is a black mark against his reputation. To make matters worse, the more nutritious skin and mucus a female eats, the bigger she grows and the greater the chances she'll change sex, turn into a male and try to take over his territory. As in their relative the Humphead Wrasse, sex is a flexible affair among cleaner wrasse.* After a few severe scoldings, the female cleaner refrains from cheating, and together the pair provides an honest service.

Beyond the complex social lives of cleaner wrasse, many other fish bear hallmarks of higher thinking, including some abilities that had been thought to be uniquely human. Guppies can count, as can sticklebacks, blind cave fish and various other fish species.† In lab tests these counting fish show their arithmetic skills by choosing between shoals of different sizes; given a choice, they generally prefer to join the largest shoal. Fish also use tools. Archerfish shoot water bullets, and tusk fish pick up clams and open them by repeatedly smashing the shells against a rock anvil. Atlantic Cod have invented a new way of feeding themselves using makeshift tools. In a Norwegian research lab a few years ago, three cod, in two different tanks, accidentally got their plastic identification tags tangled in a string that released food from an automatic feeder. All three soon figured out this was quicker than using their mouths to pull the string, because that way they had to first spit out the string before eating. These three fish perfected their

---

* There are five known species of cleaner wrasse in the *Labroides* genus, living in the Indo-Pacific region. In the Caribbean, neon gobies perform a similar role.

† Non-fish, non-human animals that can count include chimps, elephants, dogs, dolphins, pigeons, beetles and bees.

technique until they could skilfully hook themselves on the feeding string with their tags, give it a sharp tug, then turn around and swallow the food.

Another tick in the box for fish's higher cognitive abilities is the subtle way they prefer to use one side of their bodies and brains over the other.* Many individual fish prefer to look at unfamiliar objects or watch for trouble with either their left or right eye. Within a school, some fish prefer to watch their shoal mates through their left eye and accordingly spend more time in the right side of the group, while for others the reverse is true. It's possible that schools contain an optimum balance of right-lookers and left-lookers, so that collectively they watch each other and stay in formation, while at the same time keeping their outermost eyes on the look-out for predators. This asymmetry in processing and analysing information is thought to underpin our own abilities to multi-task, and it's involved in various aspects of human behaviour. Many aspects of language, for example, are generally found in the left hemisphere of a human brain.

An important aspect of intelligence is the way in which individuals interact with each other – their social intelligence. A 2012 study of captive Lemon Sharks in the Bahamas showed they learn from each other. Sharks were trained to press a target for a food reward, just as Eugenie Clark did, including the shark she took to the Japanese prince. When individuals were kept with other sharks that already knew what to do, they learned faster than those with no pre-informed tank mates.

Male cichlids from Lake Tanganyika can work out where they fit in a strict social pecking order simply by watching other males fighting. Known as Burton's Haplos, these are truculent little fish that spend a lot of their time

* This is known as cerebral lateralisation, which occurs in many different animals and varies between individuals, populations and species.

brawling over territories. Pairs of males engage in sharp skirmishes until one of them concedes defeat. The winner is easy to spot: he stands his ground and keeps the bold, black stripes between his eyes, while his opponent fades his stripes and slinks away. Researchers at Stanford University led by Logan Grosenick staged a series of fights between haplos, which they named A to E; E loses every fight; D loses to everyone except E; C only beats D and E, and so on up to A who wins every time. While pairs of fish came to blows, a third fish was allowed to watch from a safe distance in a separate, transparent compartment in the aquarium. Later, after all the fish had been given some time out and their antagonistic colours had gone back to normal, the bystander was given a choice of hanging out with one of the two fighters. Every time he chose the weaker, and hence the safer, fish. This happened even if he'd never seen the two fish actually fight each other. If his choice was between fish B and D, and he'd seen B beat C, and C beat D, he could infer that B should also beat D. He then correctly chose D as the safer male to hang out with.

Working through these logical steps is a form of deductive reasoning that some birds can apparently do, and some primates, including humans once they're four or five years old. This ability evolved in cichlids, presumably, because working out another male's rank helps them avoid getting drawn into potentially dangerous fights while maintaining the harmony of the hierarchy.

As well as fighting, many fish cooperate and help themselves by helping each other out. Back on coral reefs, predatory groupers form hunting partnerships with moray eels. When it wants some help, a grouper will hover over a spot in the reef where a moray is resting, and vigorously shimmy its body. This motion catches the eel's attention; moments later a face appears, and the pair go hunting together. The two predators make a dangerous team. Groupers prowl in open water and prey fish will try to dodge out of reach by darting into the reef. This is where the eel comes in. With its slender body, it can pursue prey through the reef's narrow cracks and nooks; either it catches the prey, or it chases it back into open water and into the grouper's waiting jaws.* A grouper and an eel working together will both get plenty to eat. Now and then the eel loses interest and wanders off. It slips away into the reef's hidden labyrinth; when it doesn't re-emerge the grouper tries to rouse its hunter partner, and once again breaks out into a shimmying dance.

At other times, while hunting alone, a grouper might adopt a different tactic if a prey fish escapes into the reef – it simply stops and waits. The grouper isn't just hoping the prey will emerge; it's also waiting for help to arrive. For up to half an hour it will hang around until another predator passes by, hopefully a moray eel or a Humphead Wrasse. Then the grouper immediately stands vertically, tail up, and rhythmically shakes its head towards the spot in the reef where the prey dived in. When an eel or a wrasse sees a grouper pointing and shaking like this, it usually wanders over to investigate. A big wrasse can't enter the reef but its powerful, extendable jaws can crunch the coral and suck out the prey from its hiding place; either that, or they'll disturb it so that once again the doomed prey leaves the reef and the grouper gets another chance to catch it.

Pointing at things is an important human trait that's considered to be a key element in language development. These expressive gestures are rare in the rest of the animal kingdom. A chimpanzee will scratch its body to indicate where it wants its companion to groom, and ravens will show each other food, perhaps as a way of forming social bonds. But until scuba-divers spent hours following the hunting, pointing groupers, these gestures were unknown among fish.

To witness and test a fish's smarts doesn't require long, involved experiments. If you have pet fish you can test their learning abilities by feeding them at one end of the tank each

---

* Groupers sometimes go hunting with octopuses, which can also slip into a reef.

morning and the opposite end in the evening, and see how long it takes until they gather at the right end ahead of feeding time – a process known as time-place learning. Usually it takes Guppies 14 days to do this (rats take almost a week longer). And the idea that Sheh Hsien, the earliest written record of the Cinderella story, had a lucky Goldfish that recognised her is not so far-fetched. Archerfish have shown that they can learn to distinguish between pictures of different human faces by shooting droplets of water at the one they've been taught to associate with food. It's highly likely, then, that pet fish learn to recognise their owners, too.

Studies of fish cognition are presenting a new view on the evolution of brains and cognition. Fish perform many behaviours previously thought to be exclusive to humans and a few other big-brained primates. This goes against a long-standing theory that primates evolved large brains to deal with the requirements of living in complex social systems. Many fish also have complex social lives and intricate behaviour despite having relatively small brains for their body size.

An alternative and often more interesting view comes not from fixating on the importance of big brains, but from asking how animal minds and cognition are affected by ecology. Brains evolve in just the same ways as other organs and behaviours: they respond and adapt to the world around them, to an animal's habitats and to other living organisms. Distantly related species may have similar mental skills because they were honed by similar environments. This could be why, for example, cichlids as well as some birds and mammals can make the deductive leap that if A beats B, and B beats C, then A must also beat C. This ability solves comparable challenges of gauging social rank. Equally, we may see closely related species with different levels of cognition because they've adapted to live in different environments. It's a Darwinian perspective of cognition that subscribes to the obvious truth that brains don't evolve in isolation; they don't float in jars lined up on shelves – they're lodged inside animals that swim, crawl and fly, hunt or graze, climb mountains or scramble through forests.

In adopting this ecological approach, the 30,000 or so species of fish become a powerful experiment in brains and thinking. Fish show how flexible brains and cognition can be, and how ecology matters.

Take, for example, the gobies living on rocky shores that memorise the world around them so they can make a quick getaway when they need to. At high tide these blotchy little fish swim around forming a mental map of local landmarks, learning the shapes of rocks and figuring out where pools will form as the water ebbs away. Then when the tide is out, if a predator looms near, the goby leaps in precisely the right direction and for the correct distance to dive straight into a nearby pool, even if it can't see where that is. When scientists remove these gobies from their homes, the fish can still remember the pool layout several weeks later. Pool-dwelling gobies are much better at learning to navigate and figuring out where they are than other gobies that live on open, flat sand. Pit these two goby species against one another in a task to learn a route through a maze, with a reward of food at the end, and the pool goby is usually the winner.

On the face of it, the sand goby looks to be the one of the pair with the lesser brain; indeed, a pool goby has a larger telencephalon, the brain region responsible for spatial learning. But if we ponder why this is, the answer comes from the sand goby's everyday surroundings. Living in a flat, featureless realm, it isn't accustomed to encountering local landmarks, so this kind of navigation simply isn't relevant; it just swims up and down the shore as the tide goes in and out. In other studies, fish grow up to be better navigators when they're reared among seaweeds and rocks rather than in plain, empty tanks; the relevant parts of their brains grow bigger, with more connections between neurons.

Throughout life, the shifting, changing environment leaves its mark on what goes on inside a fish's mind.

Bit by bit, studies like these are reframing the mainstream scientific view of fish brainpower and showing that these animals are far smarter and have much more sophisticated lives than they've long been given credit for. This raises a bigger question: are fish sentient and conscious beings?

Scientists and philosophers have historically struggled to define these concepts. Sentience involves an animal's ability to experience and feel sensations, including pleasure and pain. Consciousness is even harder to explain. As *The Blackwell Companion to Consciousness* puts it, 'anything that we are aware of at a given moment forms part of our consciousness, making conscious experience at once the most familiar and most mysterious aspect of our lives.' How, then, does this experience apply to other animals? Very broadly speaking, we can think of conscious animals as having a sense of self, and some understanding of their place in the world.

Consciousness is generally thought to be a property that emerges from higher intelligence and sentience. A key criteria for consciousness, and one we have some hope of testing for, is self-awareness – the ability to recognise and think of oneself as an individual.

For several decades, a classic way of assessing self-awareness has been the mirror test. The method involves giving an animal a mirror and watching what happens next. Many animals will initially respond to their reflection as if it were another animal; in this way experimental fish often launch a territorial attack on their reflection, assuming it's an intruder. Some animals may then inspect the mirror and look behind it, and repeatedly look into it as they begin to learn that they're looking at themselves. Chimpanzees have been seen to pick their teeth and dolphins blow bubbles in front of the mirror.

A final step in the test is for the researcher to apply a sticky coloured dot to a part of the animal's body that it can't normally see, often the forehead. Around 75 per cent of chimpanzees tested this way will look in the mirror and reach up with a hand to touch the dot. Humans start doing this aged 18 months. Primatologists interpret this as self-awareness. The chimp and young human know they're looking at themselves in the mirror; they know what they should look like and the dot on their head is something unexpected, so they poke it to see what's going on.

Only a few other species have passed the mirror test. Eurasian Magpies have looked in a mirror and scratched with their feet at a coloured dot stuck to their throat, but have ignored less conspicuous black dots on their black feathers. At the Bronx Zoo in 2006, a '2.5m- tall elephant-resistant mirror', as the study authors put it, was presented to three Asian Elephants. All three spent time in front of the mirror apparently checking out their reflections. When invisible sham marks were painted on the side of their heads all the elephants ignored them – understandably so, because there was nothing out of the ordinary for them to see. Two of the elephants also ignored a coloured mark, but one of them, a female called Happy, repeatedly reached her trunk to apparently investigate the white cross painted above her eye.

Bottlenose Dolphins and Orcas, it's often claimed, have passed the mirror test by checking out their reflections for prolonged periods and gazing at coloured dots painted on their bodies (they don't have hands, beaks or trunks that they can touch other parts of their body with, so that part of the test doesn't apply). In 2016, Csilla Ari and Dominic D'Agostino from the University of South Florida tried this on fish for the first time. They lowered a giant mirror over the side of an aquarium tank in the Bahamas, and filmed the response of two Giant Oceanic Manta Rays (with no coloured dots applied). The mantas spent a long time circling in front of the mirror, repeatedly unfurling their two cephalic lobes, the horns that channel plankton into their mouths and – like dolphins – blowing bubbles.*

Ari and D'Agostino cautiously interpreted this as a process known as contingency checking; in a similar way, you might wave a hand to check it's your distant reflection in a window. If that is what mantas are doing, then it supports the idea that they recognise themselves in the mirror and have a sense of who they are. Other researchers, though, have been highly critical of these findings and suggest the mantas were simply being social and thought they were hanging out with other mantas. The same criticisms could equally be levelled at many other mirror tests, including on marine mammals. But those studies prove to be far less controversial, perhaps because they fulfil the expectation that mammals, especially cetaceans, are intelligent, rather than challenge the deep-set belief that fish are not.

A growing appreciation of the thoughtful lives of everything from Guppies and Goldfish to mantas and cod plays into the most controversial idea surrounding fish and fish cognition: the long-standing debate over whether or not fish feel pain.

The default position for a long time has been that fish neither suffer nor feel pain. Most advocates of this position say that until and unless evidence is found to prove fish do feel pain, we should assume they don't. However, studies are beginning to provide just the kind of data to undermine this viewpoint.

In 2003, researchers from the Roslin Institute at Edinburgh University found that fish are hardwired to

* Mantas don't breath air, but it can accumulate in their gills when they're filter-feeding, which is how they're able to blow bubbles.

detect pain. The team, led by Lynne Sneddon, located in Rainbow Trout a type of nerve cells that specifically senses various noxious stimuli, including high temperatures, acid and bee venom. These cells are very similar to the sensory nerves that detect pain in mammals. Since that discovery, there's no longer been any doubt that fish have neurons dedicated to responding to stimuli that can harm them and which are associated with causing pain (albeit only in teleosts so far; equivalent receptors for noxious stimuli have not yet been found in elasmobranchs). The only remaining question is how fish perceive this sensory input.

Clues come from watching fish when they're exposed to stressful and potentially harmful situations. Numerous behavioural studies suggest that fish want such scenarios to stop – one of the hallmarks of pain. When Lynne Sneddon's team injected weak acid or bee venom into the lips of Rainbow Trout, the fish lay on the bottom of the aquarium tank and rocked from side to side, or they rubbed their lips against the side of the tank; they didn't respond like this when they were injected with harmless blanks, so it wasn't the injection itself that was upsetting them. Moreover, the trout stopped behaving this way as soon they were given a dose of morphine – one of the most potent painkillers in humans.

Pain may also dominate a fish's attention and distract it from performing other tasks (chronic or intense pain can do the same thing in humans). In 2009, Paul Ashley at the University of Liverpool led a team that tested the anti-predator responses of Rainbow Trout with and without exposure to potentially painful stimuli. When captive trout sense alarm chemicals released by damaged fish tissues, they normally swim around the aquarium tank looking for somewhere to hide.* But trout with acid-injected lips didn't

* Known as schreckstoff, these alarm chemicals are common in the otophysan fish with bony connections between their swim bladders and inner ears.

try to hide when the signal of danger rippled through their tank – they ignored the alarm chemicals and seemed to be distracted by the pain.

None of this will come as any surprise to the many biologists who consider that the detection of potentially painful stimuli and pain perception evolved hand in hand, because together these entwined processes boost an animal's chances of survival. By learning to associate dangerous situations with painful feelings, and then trying to avoid that pain, animals keep themselves out of trouble. A key driver for the formation of memories is likely to be the emotional response to pain. It's widely thought that this paired ability, to detect dangerous events and to mount an unpleasant, emotional response to them, is an ancient survival tactic that evolved early on the vertebrate lineage.

Further studies have also indicated that fish can suffer the effects of stress. Zebrafish seem to undergo emotional fevers, a rise in body temperature caused purely by stress or anxiety, something else that had previously been thought to happen only in humans (stress in the run-up to exams can give students the same physiological response as that caused by an infection). When confined to a small net, Zebrafish's temperature increased by between 2 and 4°C (3.5–7°F). What's more, farmed salmon routinely display symptoms of depression. Known as 'drop outs' in the industry, up to a quarter of the stock will have stunted growth, with these animals hanging out near the water surface where they are easy to catch. A 2016 study measured high levels of cortisol in the dejected salmon, a hormone that's commonly released in response to stress. Similar overactivity of the system that regulates cortisol levels, as well as those of the hormone serotonin, has been linked with chronic stress and depression-like states in other animals, including humans.

The case against fish feeling pain takes the stance that observed behaviours could simply be automatic reflexes that don't involve any emotional suffering. When you touch a scalding surface you feel a jolt and pull your hand away a beat before the pain kicks in; maybe fish don't get to that point of feeling pain and simply know when to withdraw themselves from danger.

Central to this argument is the contention that fish don't have a region in their brains that in humans seems to be involved in pain perception. The cerebral cortex is the outermost part of a mammal's brain. In humans, this grey matter is roughly 4mm (0.16in) thick and made up of many characteristic layers of neurons and their wiry extensions. It's folded into deep grooves and ridges, and is involved in crucial aspects of our lives, including sight, hearing, learning, feelings of suffering and stress and the perception of pain. Look inside a fish's skull and you won't find a large, mammalian-like cortex, but rather a string of small, globular beads.

No cortex: no pain. So the anti fish-pain argument runs. The premise is that fish lack the kind of complex neural architecture that allows humans to process streams of information, to extract unpleasant sensations and to know that we're hurting. The only way for other animals to feel pain is the human way. This is the stance taken in 2016 by Brian Key, a neuroscientist and prominent fish-pain sceptic from the University of Queensland in his article in the journal *Animal Sentience* entitled *Why fish do not feel pain*. Notable academics from a range of scientific disciplines published 42 written responses to Key's piece: five supported his view; two gave a neutral view that more studies were needed before making any substantive claims either way; the rest, 35, were highly critical of Key's science, reasoning and assumptions.

Among critics, neuroscientists pointed out that there's no consensus yet over just how important the cortex is for pain perception in humans, let alone what the implications are for a lack of cortex in other animals. Focusing solely on the cortex also ignores the distinct possibility that other regions of fish brains could be involved in pain perception.

The same goes for birds and various other animals that don't have a highly developed cortex, but are assumed to be sentient.

Carl Safina, Professor of Nature and Humanity at Stony Brook University in New York, uses stingray venom as testimony to fish feeling pain, in his response to Key's claims. Safina points out that stingrays, along with many other venomous species, evolved their venoms as defence against predators, including marine mammals and fish. And as we've already seen, many venomous fish evolved bright colours as a warning to predators to leave them alone or risk getting stung. For these warning colours to be effective they must be backed up by actual noxious defence (except for the mimics that skilfully imitate venomous species). Key asserts that predators need not feel that painful sting to learn to avoid it, a viewpoint that Safina rejects. 'It is nearly inconceivable that a predator would avoid the threat of a nasty sensation that it could not feel,' he writes. 'It seems logically inescapable that pain is what makes all this work.' Some animals, he points out, do seem to be immune to some of nature's stings. He describes watching sea turtles munching Lion's Mane Jellyfish and showing no signs of being stung; meanwhile, he recounts seeing a Blue Shark taking a mouthful of the same jellyfish species, then vigorously shaking its head and spitting the jellyfish out. 'The shark showed behavior consistent with pain,' Safina writes, 'the turtles do not.'

There's a lot riding on the issues of fish sentience and consciousness. Arguments are generally rooted in the scientific understanding of their pain perception and suffering – or lack thereof – but the implications resonate far beyond the boundaries of science.

This is part of a broader matter of how much of our attention, empathy and even fondness we bestow on other members of the living world. The way we treat animals and interact with them is swayed by how we think of them as sentient, intelligent beings, by the perceived simplicity or complexity of their lives. On the whole, the animals we care for most are the ones considered to be beautiful or that look back at us with thoughtful, knowing eyes: those most like us.

Since the early 19th century there's been a variety of legislation put in place to protect certain animals against pain and suffering. In 1822 the Cruel Treatment of Cattle Act was passed by the UK Parliament, banning the improper treatment of cows as well as sheep; the 1835 Cruelty to Animals Act included dogs and goats, and outlawed bear-baiting and cockfighting. Public opinion in western societies has gradually shifted towards supporting animal rights and the need to protect and look after animals in our care, from the treatment of pets and zoo animals to the design and regulation of abattoirs and the production of free-range eggs and meat.

But not all animals fall within the same legal and moral boundaries. Historically, the ethical treatment of fish has lagged far behind that of other vertebrates. Nevertheless, the science of fish cognition and sentience is catching up. Studies are chipping away at this assumption that we can get away with treating fish as lesser beings. As we've seen, scientists are devising methods to measure fish's abilities and compare them to other, more familiar animals. It's becoming clear that fish live complex, intelligent and nuanced lives, and evidence is stacking up that fish can suffer, that they get scared and they can feel pain.

Given what we now know about the lives of fish, where does that place them on our scale of empathy and ethics? How should we treat fish?

Answers to those questions are still unfolding. We're only just beginning to grapple with the potentially profound implications that come with the knowledge that fish are in fact brainy creatures. The issues are made more complicated by those emotive terms that can be difficult to grasp – sentience and consciousness, pain and suffering. Moreover, fish still face the same, age-old problems of being so different to us land-dwelling, air-breathing humans, and they live in a realm few people see and experience.

In different countries, different lines are being drawn with respect to legislation over the treatment of fish. In the UK, the Animals (Scientific Procedures) Act 1986 regulates the use of animals in scientific studies, and requires researchers to obtain a licence in order to experiment on a list of protected animals, and follow a strict Code of Practice on how to rear and handle them. The list encompasses all vertebrates plus cephalopods (once they've hatched), due to the higher cognitive powers of octopuses and their relatives. Fish are included, although specifically only once they can feed themselves independently. By contrast, all fish are excluded from equivalent legislation protecting animals in the US under the Animal Welfare Act.

There's also UK legislation protecting fish kept as pets. In 2017, a British man was convicted of cruelty to animals after he posted a video on Facebook of himself apparently swallowing a live goldfish for a bet. Officers from the Royal Society for the Prevention of Cruelty to Animals (RSPCA) saw the clip and launched an investigation. The man, along with the woman who filmed him, claimed they thought the fish was already dead. Their pleas were rejected in court and the pair were jailed for 18 weeks, ordered to do 200 hours' community service and banned from keeping fish for five years. According to the BBC news website, the convicted man said, 'I didn't think eating a fish could cause this much trouble.'

The German Welfare Act states that 'no-one may cause an animal pain, suffering or harm without good reason.' And that includes fish. Under this act, it's argued that for recreational fishers to catch and then release fish causes suffering without any 'good reason' and so this practice is illegal. All fish caught must be landed and taken home to eat (except if anglers accidentally catch undersize fish or a particular protected species). Similar legislation bans catch-and-release fishing in Switzerland. Other countries take an opposing view and encourage this approach as a conservation tool, to prevent overfishing.

Clearly there's a mixed picture, and no one way in which all fish are treated in all places. And certainly, the science of fish brains and intelligence is taking time to trickle more widely into public awareness. Resistance to a shift in attitude is fuelled, in part, by the immense vested interest in fishing industries. If fish were to be treated the same as other vertebrates, and equivalent welfare legislation introduced to the fishing and aquaculture industries as for farming, it would require radical changes to the ways wild fish are caught and to the methods used to rear them in captivity. But there's simply too much material investment, too many jobs and too much money at stake for this stance to easily take hold.

It's unrealistic, then, to expect a wholesale change in the treatment of fish any time soon. Perhaps, though, we can hope for a shift of public opinion in favour of fish, one that allows them greater respect and appreciation. Already we've moved on from myths of a Goldfish's seven-second memory and thrown out many defunct beliefs about the fish's dull-witted lives. It turns out they do have all those fundamental abilities it was variously assumed they lack: they can think, learn and remember, they can see colours, they can hear and sing. And fish have countless other curious and unexpected talents: they cast electric beams around them to hunt and find their way; they send out covert messages by manipulating light and colour; they sculpt giant shapes in the sand; and they use inbuilt magnetic compasses to swim across the oceans and back. But still there's a long way to go to redirect the common view that fish are inferior creatures, that they're wholly different from dogs, horses, cats, birds and all the animals that have been regarded as loyal parts of human life for thousands of years.

There are many ways we can redress this balance and close the conceptual gap between fish and other animals. We can pay attention to the fish that we eat, and keep as pets, and ask questions about where they came from and how they were caught or farmed. We can advocate for continued research into the lives of fish, to learn yet more of their biology – and pay good attention to the results of those studies – and we can learn more about how human activities affect them as individuals, populations and species. And you can cultivate your own, personal connections, and get to know these animals better by going into their aquatic world and giving yourself over to the joys of watching fish.

Epilogue

On a Tuesday afternoon, I leave my house and cycle across the city to go in search of fish. I pedal through fenland that smells of warming cow pats and over an iron bridge, like a miniature railway, that's hung with silk from spiderlings' early summer flights. I pass the lone lamppost in the meadow where 80 years ago people came to skate in frozen winters, but now cut off from the river's flood it's home to orchids and grass snakes. Down a green tunnel, over the rattling ridges of a cattle grid and into the bright open path along the riverside.

Two red camping chairs with recumbent anglers are hopeful signs that I'll find something. I prop my bike against a tree and find a spot to sit and look. The water is cloudy brown, the surface reflects blue sky and puffs of cloud, and I see nothing below. Overhead a tern follows the river's path, one way and then back again, squeaking now and then like a dog's chewy toy. Its pointed wings are swept back, and a black cap hides its eyes that look down for food. It spots something and drops with a splash, then flaps off, swallowing. I spy the tern's target as a shoal of fish fry sweep into a shallow recess in the riverbank next to me. They have big, dark eyes and transparent bodies. A couple of them are bold and lead at the front, while others trail behind, as the shoal inspects water-filled, muddy footprints in the riverbed, which to them must be like giant craters.

Their motion seems hesitant and alert. Swim, swim, pause. Swim, swim, pause.

As I lean in to get a closer look they flit away and gather around a frond of river weed, apparently sensitive to my presence. They are far less troubled by a kayaker paddling past; maybe their chief enemies are aerial and bankside predators, not splashing waterborne beasts. When the water flattens again, the fish fry rise to the surface and kiss wrinkled circles that expand across the water's skin.

I walk beside the river, past students with disposable barbecues scorching squares in the grass, past a man in swimming trunks sitting quietly under a low-hanging tree. A sunbathing couple listen to tinny tunes, which male damselflies dance to, trying to impress their mates with shining, sapphire bodies and pigmented wings that flutter like butterflies.

Then, in a clearing in the murky water, I see two fish silhouettes hanging in the current, at least a hand-span in size, tails pointing downstream towards the city. I find a gap in the thistles and nettles, scramble down the bank and step shin-deep into soft mud. The riverbed slopes sharply and I quickly get out of my depth, lunging into water cold enough to steal my first three breaths. My pasty white legs appear whisky-brown through the water and I float for a moment, fiddling with my dive mask, spitting and rubbing to stop it fogging.

It feels strange to sink so low, without the boost of saltwater I'm accustomed to. I'm also used to being able to see further than this. Thick water pushes against the glass of my mask and I wonder if I'll see anything at all. In the river's main channel a fish would have to swim right past my nose for me to spot it.

I look across the water surface that's flecked with fluff shed by the surrounding willow trees and decide to try my luck along the far bank. I swim over a silky bed of waterweed, which wraps around my legs, and I push my way into the overhanging vegetation, disturbing a hidden moorhen and her chick that clamber around at my eye level. A punt comes past, a long way from the city-centre tours, and the man poling it along says to his reclining passenger, 'There's a swimmer looking at things.' I reply with a cheerful 'Hello!' and return to my underwater search.

Holding as much air in my lungs as possible to stay afloat, I do my best to keep my feet away from the soft riverbed that will ruin my view. In this quiet backwater, between the submerged branches and roots, the mud has settled down and I can now see at least an arm's length ahead.

The scene reminds me of snorkelling in a mangrove forest in Madagascar. Twice a day, at high tide the amphibious trees there are flooded, and an aquatic ecosystem swiftly assembles around the looping roots and knuckled trunks. Until then I'd assumed mangroves were essentially all mud and there would be little point looking beneath the waterline. But I marvelled at the water's clarity, at the silver fish weaving through the trees and the sweeping schools of fry that split and reformed around me. Here, my river view is not quite as clear and it's far less busy than the tropical forest, but I have the same sense of this liminal space, connecting land and water.

I wait patiently for something to come by, and begin to worry that all the fish are avoiding me. Then, at last, one of them dares to come near and hovers in view, treading the water with undulating pectoral fins. It has large, overlapping silver scales and red fins, including a deeply forked tail. It's a Roach, a species that ranges across Europe from the Pyrenees to Siberia, but this one, just for a moment, is all mine. With a red-rimmed eye it watches me and I watch back, holding as still as I can in case I hurry it away. I count five sips of water it sucks in that mark the brief seconds we spend together. Then, with a body twitch and a flick of its tail, the Roach slides out of sight, leaving me alone once more, floating in the river.

Appendix: Illustration species list

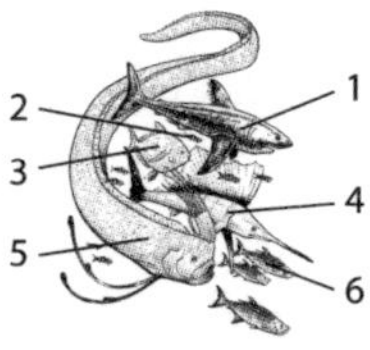

Prologue

1 Salmon Shark *Lamna ditropis*
2 Pilotfish *Naucrates ductor*
3 Opah *Lampris guttatus*
4 Atlantic Sailfish *Istiophorus albicans*
5 Giant Oarfish *Regalecus glesne*
6 Atlantic Tarpon *Megalops atlanticus*

Chapter 1: Ichthyo-curiosities

1 Monkfish from Belon's *De Aquatilibus*

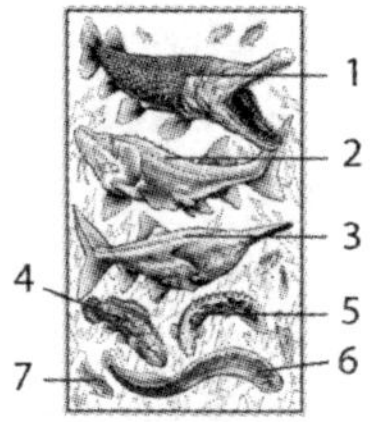

Chapter 2: A view from the deep

1 Alligator Gar *Atractosteus spatula*
2 Sturgeon (Acipenseridae)
3 Spoonbill Paddlefish *Polyodon spathula*
4 Bowfin *Amia calva*
5 Bichir (Polypteridae)
6 Lamprey (Petromyzontiformes)
7 Cichlid (Cichlidae)

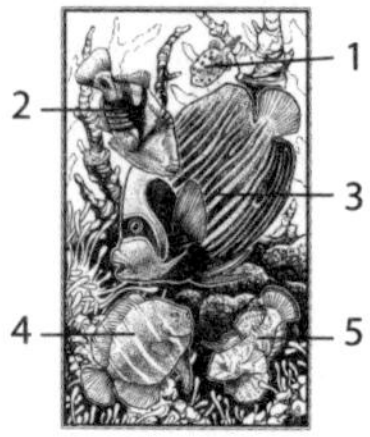

Chapter 3: Outrageous acts of colour

1 Yellow Boxfish *Ostracion cubicus*
2 Picasso Triggerfish *Rhinecanthus aculeatus*
3 Emperor Angelfish *Pomacanthus imperator*
4 Peppermint Angelfish *Centropyge boylei*
5 Mandarinfish *Synchiropus splendidus*

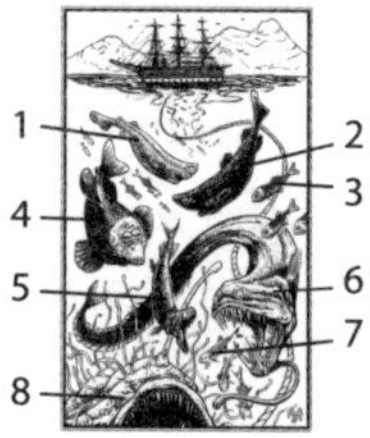

Chapter 4: Illuminations

1 Cookie Cutter Shark *Isistius braziliensis*
2 Lanternshark *Etmopterus*
3 Lanternfish (Myctophidae)
4 Pacific Barreleye *Macropinna microstoma*
5 Stoplight Loosejaw *Malacosteus niger*
6 Dragonfish (Stomiidae)
7 Bristlemouths (Gonostomatidae)
8 Illuminated Netdevil *Linophryne arborifera*

Chapter 5: Anatomy of a shoal
1 Cow rays *Rhinoptera*
2 Humphed Wrasse *Cheilinus undulatus*
3 Moorish Idols *Zanclus cornutus*

Chapter 6: Fish Food
1 River Hatchetfish *Gasteropelecus sternicla*
2 Black Spot Piranhas *Pygocentrus cariba*
3 Electric Eel *Electrophorus electricus*
4 Cardinal Tetras *Paracheirodon axelrodi*

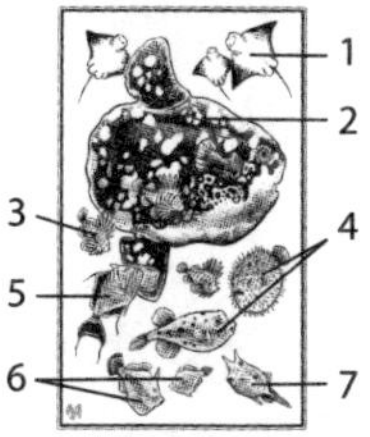

Chapter 7: Toxic fish
1 Stingrays (Dasyatidae)
2 Ocean Sunfish *Mola mola*
3 Lionfish *Pterois*
4 Blackspotted Pufferfish *Arothron nigropunctatus*
5 Triggerfish (Balistidae)
6 Fringed Filefish *Monocanthus ciliatus*
7 Cowfish *Latoria*

Chapter 8: How fish used to be
1 *Bothriolepis*
2 *Materpiscis*
3 *Doryaspis*
4 *Dunkleosteus*
5 *Harpagofututor*
6 *Stethacanthus*
7 *Edestus*
8 *Helicoprion*
9 *Leedsichthys*

Chapter 9: Fish symphonies
Haddock *Melanogrammus aeglefinus*

Chapter 10: (Re)thinking fish
1 Surgeonfish (Acanthuridae)
2 Roving Coral Grouper *Plectropomus pessuliferus*
3 Giant Moray Eel *Gymnothorax javanicus*
4 Bluestreak Cleaner Wrasse *Labroides dimidiatus*

Glossary

Acanthodians Extinct spiny sharks, thought to be the direct ancestors of living chondrichthians.

Actinopterygians The ray-finned fish, including teleosts, bowfins, gars, bichirs and sturgeon.

Barbel A fleshy projection, usually somewhere on the head.

Caudal fin A fish's tail.

Caudal peduncle The region at the base of a fish's tail.

Chondrichthyes Class containing sharks, rays, skates and chimaeras.

Chordates Phylum of animals, including fish and all other vertebrates, along with sea squirts.

Chromatophore Pigmented skin cell.

Claspers Modified fins of male sharks, rays and skates (and extinct placoderms) that deliver sperm during mating.

Conodonts Extinct group of jawless fish.

Dorsal fin Fin on a fish's back.

Elasmobranchs Sub-class of fish containing sharks, rays and skates.

Galeaspids Extinct group of jawless fish.

Ichthyo-curiosities A word I made up to hint at the many wonders of fish.

Ichthyology The study of fish.

Iridophore Skin cell that produces structural colours, and gives fish their shine.

Lateral line System of pores and tubes that let fish detect changes in water pressure.

Osteostracans Extinct group of jawless fish.

Otolith Dense structure, made mostly of calcium carbonate, found in the inner ear of fish, used for hearing and balance.

Otophysans Fish with a bony connection between their inner ear and swim bladder that boosts their hearing, including 60 per cent of all freshwater fish.

Pelagic Anything belonging to the realm of the open sea.

Pharyngeal teeth Second set of teeth located deep in a fish's throat.

Placoderms Extinct group of fish, the first to evolve jaws.

Sarcopterygians The lobe-finned fish, including lungfish, coelacanths and extinct tetrapodomorphs.

Swim bladder The internal gas balloon, adapted from an ancestral lung, which fish use as a flotation device and to boost their hearing.

Teleosts The latest group of bony fish to split off on the fish evolutionary tree, accounting for 96 per cent of known fish species.

Tetrapodomorphs Extinct group of lobe-finned fish that gave rise to the tetrapods (four-limbed vertebrates).

Thelodonts Extinct group of jawless fish.

Select bibliography and Notes

Fish common names, sizes and ages I have taken from *Fishbase* or the *Encyclopedia of Life* websites.

* Denotes open access papers, freely available online. Find web links to these papers at www.helenscales.com/fishscience

Prologue – The wandering ichthyologist

Figures on annual global fish catches come from the UK fish welfare organisation fishcount.org.uk.

Ichthyo-curiosities

Jordan, D. S. 1902. The history of ichthyology. *Science* 16: 241–258.

Kusukawa, S. 2000. The Historia Piscium (1686). *Notes Rec. R. Soc. Lond.* 54: 179–197.

Nelson, J. S., Grande, T. C. & Wilson, M.V.H. 2016. *Fishes of the World*. John Wiley & Sons, Hoboken, New Jersey.

'Destitute of feet' and 'lanky fish' from D'Arcy Wentworth Thompson's 1910 translation of Aristotle's *History of Animals*.

A view from the deep – introducing the fish

Helfman, G., Collette, B.B., Facey, D.E. & Bowen, B.E. 2009. *The Diversity of Fishes: Biology, Evolution and Ecology*. John Wiley & Sons, Hoboken, New Jersey.

Nielsen, J. et al. 2016. Eye lens radiocarbon reveals centuries of longevity in the Greenland shark (*Somniosus microcephalus*). *Science* 353: 702704.

Standen, E., Du T.Y. & Larsson, H.C.E. 2014. Development plasticity and the origin of tetrapods. *Nature* 513: 54–58.

Takezaki, N. & Nishihara, H. 2017. Support for lungfish as the closest relative of tetrapods by using slowly evolving ray-finned fish as the outgroup. *Genome Biology and Evolution* 9: 93–101.*

Outrageous acts of colour

Endler, J.A. 1980. Natural selection on colour patterns in *Poecilia reticulata*. *Evolution* 34: 76–91.*

Reznick, D.N., Shaw. F.H., Rodd, F.H. & Shaw, R.G. 1997. Evaluation of the rate of evolution in natural populations of guppies (*Poecilia reticulata*). *Science* 275: 1934–1937.

Seehausen, O., van Alphen, J.J.M. & Witte, F. 1997. Cichlid fish diversity threatened by eutrophication that curbs sexual selection. *Science* 277: 1808–1811.

Illuminations

Anthes, N., Theobald, J., Gerlach, T., Meadows, M. G. & Michiels, N.K. 2016. Diversity and ecological correlates of red fluorescence in marine fishes. *Frontiers in Ecology and Evolution* Volume 4: Article 126.*

Davis, M.P., Sparks, J.S. & Smith, W.L. 2016. Repeated and widespread evolution of bioluminescence in marine fishes. *PLoS ONE* 11: e0155154.*

Douglas, R.H., Partridge, J.C., Dulai, K.S., Hunt, D.M., Mullineaux, C.W. & Hynninen, P.H. 1999. Enhanced retinal longwave sensitivity using a chlorophyll-derived photosensitiser in *Malacosteus niger*, a deep-sea dragon fish with far red bioluminescence. *Vision Research* 39: 2817–2832.*

Michiels, N.K. et al. 2008. Red fluorescence in reef fish: A novel signalling mechanism? *BMC Ecology* 8:16. *

Sparks, J.S. et al. 2014. The covert world of fish biofluorescence: A phylogenetically widespread and phenotypically variable phenomenon. *PLoS ONE* 9:e83259.*

Anatomy of a shoal

Barthem, R.B. et al. 2017. Goliath catfish spawning in the far western Amazon confirmed by the distribution of mature adults, drifting larvae and migrating juveniles. *Scientific Reports* 7: 41784.*

Doherty, P.D. et al. 2017. Long-term satellite tracking reveals variable seasonal migration strategies of basking sharks in the north-east Atlantic. *Scientific Reports* 7: 42837.*

Naisbett-Jones, L.C. et al. 2017. A magnetic map leads juvenile European Eels to the Gulf Stream. *Current Biology* 27: 1236–1240.*

Svendsen, M.B.S. et al. 2016. Maximum swimming speeds of sailfish and three other large marine predatory fish species based on muscle contraction time and stride length: a myth revisited. *Biology Open* 5: 1415–1419.*

Payne, N.L. et al. 2016. Great hammerhead sharks swim on their side to reduce transport costs. *Nature Communications* 7: 12889.*

Fish food

Allgeier, J.E., Valdivia, A. Cox, C & Layman C.A. 2016. Fishing down nutrients on coral reefs. *Nature Communications* 7: 12461.*

Bellwood, D.R., Goatley, C.H.R., Bellwood, O., Delbarre, D.J. & Friedman M. 2015. The rise of jaw protrusion in spiny-rayed fishes closes the gap on elusive prey. *Current Biology* 25: 2696–2700.*

Lissmann, H.W. 1958. On the Function and Evolution of Electric Organs in Fish. *Journal of Experimental Biology* 35: 156–191.*

Perry, C.T., Kench. P.S., O'Leary, M.J., Morgan, K.M. & Januchowski-Hartley, F. 2015. Linking reef ecology to island building: Parrotfish identified as major producers of island-building sediment in the Maldives. *Geology* 43: 503–506.*

Vailati, A., Zinnato, L. & Cerbino, R. 2012. How archer fish achieve a powerful impact: hydrodynamic instability of a pulsed jet in *Toxotes jaculatrix*. PLoS ONE 7: e47867.*

White, W. T. et al. 2017. Phylogeny of the manta and devilrays (Chondrichthyes: mobulidae), with an updated taxonomic arrangement for the family. *Zoological Journal of the Linnean Society* 182: 50–57.

At the time I encountered mantas in Fiji 2016, they were still considered to be members of the *Manta* genus. In 2017, taxonomists reshuffled mantas and their relatives, the devil rays, and put them all into a single genus, *Mobula*, with eight species. But we can still call them manta rays even though their genus has been banished.

Toxic fish

Casewell, N.R. et al. 2017. The evolution of fangs, venom, and mimicry systems in blenny fishes. *Current Biology* 27: 1–8.

Clark, E. 1953. *The lady and the spear.* Harper, New York.

Clark, E. 1969. *The lady and the sharks*. Harper & Row, New York.

Clark, E., Nelson, D.R. & Dreyer, R. 2015. Nesting sites and behavior of the deep water triggerfish *Canthidermis maculata* (Balistidae) in the Solomon Islands and Thailand. *Aqua International Journal of Ichthyology* 21: 1–38.

Inglis, D. 2010. The zombie from myth to reality: Wade Davis, academic scandal and the limits of the real. *Scripted* 7: 351–369.*

Quotes from Genie on her deep dives later in life come from her *Washington Post* obituary by Juliet Eilperin, 26 February 2015.

How fish used to be

Benton, M. 2014. *Vertebrate palaeontology.* John Wiley & Sons, Hoboken, New Jersey.

Ferrón, H.G. & Botella, H. 2017. Squamation and ecology of thelodonts. *PLoS ONE* 12: e0172781.*

Long, J.A., Trinajstic, K., Young. G.C. & Seden, T. 2008. Live birth in the Devonian period. *Nature* 453: 650–6522.

Sibert, E.C. & Norris, R.D. 2015. New Age of Fishes initiated by the Cretaceous – Paleogene mass extinction. *PNAS* 112: 8537–8542.*

Fish symphonies

Ruppé, L., Clément, G., Herrel, A., Ballesta, L., Décamps, T., Kéver, L. & Permentier, E. 2015. Environmental constraints drive the partitioning of the soundscape in fishes. *PNAS* 112: 6092–6097.*

Radford, C.A., Stanley, J.A., Simpon, S.A. & Jeffs, G.A. 2011. Juvenile coral reef fish use sound to locate habitats. *Coral Reefs* 30: 295–305.

Wilson, B., Batty, R.S. & Dill, L.M. 2003. Pacific and Atlantic herring produce burst pulse sounds. *Proc. Roy. Soc. London* B 271: S95–S97.

Marie Fish's sound recordings are available to listen to online from the Macaulay Library at the Cornell Lab of Ornithology.

(Re)thinking fish

Brown, C., Laland, K. & Krause, J. 2011. *Fish cognition and behavior.* Wiley-Blackwell, Hoboken.

Brown, C. 2016. Fish pain: an inconvenient truth. *Animal Sentience* 3(32).*

Grutter, A.S. 2004. Cleaner Fish Use Tactile dancing behavior as a preconflict management strategy. *Current Biology* 14: 1080–1083.*

Key, B. 2016. Why fish do not feel pain. *Animal Sentience* 3(1).*

Pinto, A., Oates, J., Grutter, A. & Bshary, R. 2011. Cleaner wrasses *Labroides dimidiatus* are more cooperative in the presence of an audience. *Current Biology* 21: 1140–1144.*

Sneddon, L.U., Braithwaite, V.A. & Gentle, M.J. 2003. Do fishes have nociceptors? Evidence for the evolution of a vertebrate sensory system. *Proc. Roy. Soc.* B 270: 1115–1121.

Sources of fish stories

Sedna the sea goddess Adapted from various sources including Laugrand, F. & Oosten, J. 2009. *Sedna in Inuit shamanism and art in eastern Arctic.* The University of Alaska Press, Fairbanks, Alaska.

How the flounder lost its smile Adapted from Morris, S. 1911. *Manx Fairy tales.* David Nutt, London.

The salmon of knowledge Adapted from various sources including Rolleston, T.W. 1910 *The High Deeds of Finn and other Bardic Romances of Ancient Ireland*, G. G. Harrap & Co., London.

O Namazu Adapted from various sources including Volker, T. 1975. *The animal in Far Eastern Art.* E.J. Brill, Leiden.

Osiris and the elephantfish This version of the Osiris myth was told to me by Egyptologist Dr Meghan Strong.

Vatnagedda Adapted from Davidsson, O. 1900. Icelandic fish folklore. *Scottish Review* 36: 312–331.

Chipfalamfula Adapted from Knappert, J. 1977. *Bantu myths and other tales.* E. J. Brill, Leiden.

The doctor of the sea Adapted from various sources including Stothard, P. January 25 2005. Islam's missing scientists. *The Times Literary Supplement* digital edition.

The fish and the golden shoe Adapted from Jameson, R.D. 1982. *Cinderella in China.* In *Cinderella, a case book.* Ed. Dundes, A. University of Wisconsin Press. Madison, Wisconsin.

Acknowledgements

For letting me build this aquarium of a book and fill it with so many fish, thank you to everyone at Bloomsbury, especially Anna MacDiarmid and Jim Martin in London who have quietly let me get on with things but have always been there when I need them. Aaron John Gregory, thank you for so brilliantly depicting all the beautiful fish that swim through these pages alongside my words. AJ, you know this book was really just an excuse to work with you again and to tap into our shared fishy obsessions.

To everyone I've ever been fish-watching with thank you, for everything you've pointed out to me and all those moments we've shared. Special thanks to the good folks at the Mole Valley Sub Aqua Club, especially Helena Egerton (former Chicken Number 2, or was it 1?), who helped get me started in this fish-watching business, and to Alice El Kilany for many of my earliest and most fun dives, in Belize, Australia and the Philippines.

From my trips that feature specifically in this book, I want to extend my heartfelt thanks to Jess Cramp and Kirby Morejohn for the incredible free diving in Rarotonga, to Lori and Pat Colin for sharing fish stories in Palau, and Sarah Frias-Torres for taking me to see Goliath Groupers in Florida. In Fiji, special thanks to Ian Campbell at WWF Pacific for supporting my manta-watching trip, along with Jessamy Ashton for so kindly giving us a place to stay and a car to drive (and for so graciously eating the curry I cooked in the dogs' dish). My huge thanks also to Heather and Dan Bowling, and all the team at Barefoot Manta Resort in the Yasawas for your wonderfully warm welcome, for introducing me to those beautiful mantas and for arranging the fluorescent night dive. In Borneo, huge thanks to Anna Petherick for being such a fun research assistant; I'm so glad spending all that time with me in windowless hotel rooms and poking around smelly fish markets didn't put you off being pals with me for so many years since then. You are fab. In Florida, thank you to everyone who welcomed me at Mote Marine Laboratory and especially to Hayley Rutger for arranging my lunch date with Eugenie Clark.

For their indispensable help with my desk-bound research, thank you to Nico Michiels, Ken McNamara and Culum Brown. For reading chapter drafts and putting up with so much fish talk, my huge thanks to everyone at Neuwrite London, especially Emma Bryce, Roma Agrawal, Vanessa Potter, Alice Gregory, Curtis Asante, Christine Dixon, Grace Lindsay and Ed Bracey. Liam Drew, thank you for not pushing the issue of which are better – fish or mammals – and for helping make many parts of this book better than they would have been.

And to all my family and friends, thank you, mainly for being so tolerant of my absence over the last two years, first when I disappeared to chase after fish and then when I came back and shut myself away to write about them. My love and gratitude to my parents, Di and Tom Hendry, especially for standing beside that freezing cold, flooded gravel pit in Leicestershire all those years ago and watching me vanish into the gloomy water, and for supporting me on every watery adventure I've been on since. You have always cheered me on, in so many ways. And to stick to literary convention, Ivan, I thank you last of all. Once again you've sustained me through another book, as my tireless reader, my storyteller, my quality controller and my fine-wine provider. And you are there, all the way through this book, in the background of so many scenes, when we were watching the same fish and floating through the same seas. Where are we going next?

Index